T0135805

Modeling with Ambient B-Splines

Ambient B-Splines: Theorie und Modellierung
Zur Erlangung des Grades eines Doktors der Naturwissenschaften (Dr. rer. nat.) genehmigte Dissertation von Dipl.-Math. Nicole Lehmann aus Wiesbaden
September 2013 — Darmstadt — D 17

Fachbereich Mathematik
Geometrie und Approximation

Modeling with Ambient B-Splines
Ambient B-Splines: Theorie und Modellierung

Genehmigte Dissertation von Dipl.-Math. Nicole Lehmann aus Wiesbaden

1. Gutachten: Prof. Dr. Ulrich Reif
2. Gutachten: Prof. Dr. Klaus Höllig

Tag der Einreichung: 13. September 2013
Tag der Prüfung: 7. November 2013

Darmstadt D 17

Bibliografische Information der Deutschen Nationalbibliothek

Die Deutsche Nationalbibliothek verzeichnet diese Publikation in der Deutschen Nationalbibliografie; detaillierte bibliografische Daten sind im Internet über http://dnb.d-nb.de abrufbar.

ISBN 978-3-8325-XXXX

Logos Verlag Berlin GmbH
Comeniushof, Gubener Str. 47,
10243 Berlin
Tel.: +49 (0)30 42 85 10 90
Fax: +49 (0)30 42 85 10 92
INTERNET: http://www.logos-verlag.de

Contents

Abstract

The present thesis introduces a new approach for the generation of C^k-approximants of functions defined on closed submanifolds for arbitrary $k \in \mathbb{N}$. In case a function on a surface resembles the three coordinates of a topologically equivalent surface in $\mathbb{R}^3$, we even obtain C^k-approximants of closed surfaces of arbitrary topology. The key idea of our method is a constant extension of the target function into the submanifold's ambient space.

In case the *reference* submanifolds are embedded and C^2, the usage of standard tensor product B-splines for the approximation of the extended function is straightforward. We obtain a C^k-approximation of the target function by restricting the approximant to the reference submanifold. We illustrate our method by an easy example in $\mathbb{R}^2$ and verify its practicality by application-oriented examples in $\mathbb{R}^3$. The first treats the approximation of the geoid, an important reference magnitude within geodesy and geophysics. The second and third example treat the approximation of geometric models. The usage of B-splines not only guarantees full approximation power but also allows a canonical access to adaptive refinement strategies. We elaborate on two hierarchical techniques and successfully apply them to the introduced examples.

Concerning the modeling of surfaces by the new approach, we derive numerically robust formulas for the determination of normal vectors and curvature information of a target surface which only need the spline approximant as well as the normal vectors and curvature information of the reference surface.

Zusammenfassung

Die vorliegende Arbeit stellt eine neue Methode zur C^k-Approximation von Funktionen auf Untermannigfaltigkeiten für beliebige $k \in \mathbb{N}$ vor. Falls eine betrachtete Zielfunktion auf einer Fläche die einzelnen Koordinaten einer topologisch äquivalenten Fläche in $\mathbb{R}^3$ darstellt, haben wir automatisch die Möglichkeit, geschlossene Flächen mit beliebiger Differenzierbarkeitsordnung zu modellieren. Dabei spielt die Erweiterung der zu approximierenden Funktion in den umgebenden Raum der Untermannigfaltigkeit eine Schlüsselrolle.

Sofern die sogenannten *Referenzuntermannigfaltigkeiten* eingebettet und C^2 sind, können wir für die Approximation der erweiterten Funktion klassische Tensorprodukt B-Splines verwenden. Die Approximation der Zielfunktion erhalten wir durch die Einschränkung des Approximanden auf die Referenzuntermannigfaltigkeit. Wir veranschaulichen unsere Methode anhand eines einfachen Beispiels in $\mathbb{R}^2$ und überprüfen ihre Funktionalität mit Hilfe anwendungsbezogener Beispiele in $\mathbb{R}^3$. Im ersten Beispiel approximieren wir das Geoid, eine wichtige Bezugsgröße innerhalb der Geodäsie und Geophysik. Im zweiten und dritten Beispiel untersuchen wir die Approximation geometrischer Objekte. Der Einsatz von B-Splines garantiert maximale Approximationskraft sowie einen einfachen Zugang zu lokalen Verfeinerungsstrategien. Wir stellen zwei hierarchische Ansätze vor und verifizieren deren Ergebnisse anhand der bereits betrachteten Beispiele.

Da man sich beim Modellieren von Flächen häufig für Normalenvektoren und Krümmungsinformationen interessiert, leiten wir außerdem numerisch robuste Formeln her, die uns diese Zielgrößen in Abhängigkeit des Splineapproximanden sowie der Normalen bzw. Krümmungsinformationen auf der Referenzfläche liefern.

1 Introduction

Computer graphics has become a broad and active research field since its dawn in the nineteen-fifties and sixties. At the beginning, it was mainly promoted by research of the military, universities as well as automotive and aerospace industries in the United States, see [Mac78] for more details. One important research area concerns the development of mathematical representations of smooth surfaces of arbitrary topology. While there are already many approaches for this, see Chapter 2, they all have in common that the generation of parametrizations of arbitrary differentiability requires elaborate work.

The present thesis introduces a new modeling approach for closed surfaces with which we may generate C^k-parametrizations of any differentiability $k \in \mathbb{N}$. It is actually a by-product of a new ansatz for the more general problem of approximating functions defined on submanifolds because the function to be approximated may represent surface coordinates in $\mathbb{R}^3$ which have been mapped to a topologically equivalent other so-called *reference* surface.

There are usually two different options of managing approximations of functions on curved domains: one either directly defines rather complicated function spaces on the submanifold itself, for example with the help of a triangulation on the sphere, cf. Section 2.2, or one works with charts and transition functions if possible and transforms the approximation problem into a manageable domain in $\mathbb{R}^{\bar{d}}$, if the underlying function is defined on a $\bar{d}$-dimensional submanifold, see Section 3.1. Both approaches have the drawback that for each individual problem one either needs to construct the function spaces on the submanifolds or the charts and transition functions, i.e., there is no universal set of functions which can be applied for arbitrary submanifolds.

We overcome the drawback of defining special functions for each problem individually by simply transforming the approximation problem into the submanifold's ambient space. First, we extend the function constantly into normal direction. Through a suitable choice of the extension domain we may use standard tensor product B-splines in $\mathbb{R}^d$ to approximate the extended function. The restriction of the approximant to the primary submanifold provides an approximation of the target function. By choosing coordinate-wise B-splines of order n, we get approximants of differential order $n-2$. Reconsidering the special case when the target function itself represents a surface in $\mathbb{R}^3$, we are able to produce C^k-approximants of surfaces of arbitrary topology and arbitrary $k \in \mathbb{N}$ which consist of one single function. The latter is remarkable: although tensor product splines are a widespread standard tool for parametric surface modeling because of their simplicity as well as their flexibility, the tensor product structure usually fails to model non-toroidal surfaces without singularities since they induce a quad structure on the domain, see Section 2.1. Our approach does not suffer a reduction of differentiability because of the additional dimension in the preimage.

The ambient B-spline method guarantees full approximation power which we confirm with the help of examples in $\mathbb{R}^2$ and $\mathbb{R}^3$. The mapping of a target surface to a reference surface only ensures the correct topological classification. We usually face an uneven distribution of geometric information and therefore suggest the usage of two adaptive refinement techniques in order to adequately approximate high frequency parts of the function. The basic necessity of the latter is in favor of using tensor product B-splines, too, because B-splines intrinsically imply an easy access to refinement strategies. Whereas the first adaptive approach is a heuristic and suitable for irregular distributions of details, the more complex second method is applied for examples where surface features or high frequency parts are only within a distinct region of the reference submanifold.

The ambient B-spline method establishes a possibility to parametrize a discrete surface by a single spline function. Extensive research deals with the challenging problem of estimating normal vectors and curvature information of surface meshes. Of course, with the composition of the spline approximant and given charts of the underlying reference submanifold we would have simple access to these quantities with standard differential geometric tools. However, we also introduce numerically robust formulas for which explicit charts of the reference submanifold are not needed if normal vectors and curvature are already given. In contrast to most discrete estimation schemes, our approximation of normals and curvature information then indeed belongs to an existing surface which approximates the target mesh.

This thesis is organized as follows:

In Chapter 2 we provide an overview of existing surface modeling techniques. Furthermore, we elaborate on the less active research field of approximating functions on surfaces or more generally on submanifolds and introduce significant works which have influenced the present thesis.

Chapter 3 introduces the *ambient B-spline method* while focusing on theoretical aspects. The method provides the possibility of using tensor product B-splines for the approximation of functions and therewith of surfaces defined on submanifolds without generating singularities though being independent of the underlying topology. We briefly investigate on its approximation power and refer to current research work of Odathuparambil, Prasiswa and Reif who provide a proof for the full approximation power of the ambient B-spline method. We mention possible difficulties within the spline approximation on general domains in $\mathbb{R}^d$.

Chapter 4 is devoted to a concrete description of the implementation of the ambient method including the introduction of notations and an illustration via a simple example in $\mathbb{R}^2$. We introduce basic preliminaries of the used B-spline functions and investigate on first results of three different examples which are all application-oriented. Additionally, first results reveal the need for a local refinement strategy.

In Chapter 5 we elaborate on two alternatives for an adaptive B-spline approximation strategy. The first is a heuristic which is easily applicable although it does not produce a best approximation in the least-squares sense. The second option lacks some careful definitions of the approximation domain as well as some investment concerning the implementation.

Chapter 6 readdresses the examples of Chapter 4 with the adaptive approaches introduced before. All examples reveal that different types of functions can be approximated well if we apply one of the two considered hierarchical strategies.

Chapter 7 recapitulates work of [LR12] in which we derive formulas for curvature quantities and normal vectors of hypersurfaces which are modeled by a deformation of a reference hypersurface. The last theorem of the chapter provides a numerically stable curvature evaluation when surfaces are modeled by the ambient method. Additionally, we apply the derived formulas to two surface examples. Since these examples do not provide reference data with which we could compare our results, we additionally investigate an algebraic surface for which we can compare the approximated quantities to the exact ones.

Chapter 8 briefly describes two considerations we hoped would improve the surface modeling. Both ideas fail to meet our expectations and are therefore not elaborated in detail.

The thesis closes by a summary of the main results in Chapter 9 and provides a discussion on open issues and promising future work.

Acknowledgments

First of all I would like to thank Prof. Dr. Ulrich Reif for his support and confidence during the development of this thesis. He suggested the topic of the present research work. Besides the fruitful mathematical discussions, I greatly appreciated the helpful implementation hints and various presentation advices.

I am also very grateful to Prof. Dr. Klaus Höllig for being the coreferee for this thesis.

I thank the *Fraunhofer IGD* for the financial support via the *Visual Computing Initiative Hessen*.

A big thank you goes to Dr. Karoline Disser, Christian Eisele, Tobias Ewald, Adrian Krion, Karin Lehmann, Sonja Odathuparambil and Dr. Julia Plehnert for proofreading and various helpful comments and remarks.

It has been a privilege to be part of a cooperative environment within the working group *Geometrie und Approximation*. Besides my advisor, I want to mention explicitly two further persons: thanks to Mrs. Drexler for supporting my colleagues and me in our daily work and to Dr. Julia Plehnert who became a dear friend and was always available for a kind word if encouragement was necessary.

2 Related Work

In this chapter, we summarize existing approaches for surface modeling and approximation of functions on submanifolds. As explained below, these two problems are closely related. We derive a surface modeling technique from the general procedure of approximation functions on submanifolds (surfaces), see Section 4.3. However, some works concerning surface modeling generate approximation techniques for functions on manifolds as a byproduct. We therefore start with an overview of surface modeling techniques because it turns out to be the more active research field.

2.1 Surface modeling

The reconstruction and representation of real-world objects play a key role within computer graphics. For instance, 3D modeling is extensively used in architecture, aerospace, automotive as well as plastic industry, biomedical engineering, earth science or animation industry. One can distinguish between solid or shell models, respectively. Solid models describe the volume of objects or can be defined as surfaces with their interiors, cf. [BW97], whereas shells (*surfaces*) only describe their boundaries. Solid models are not considered in this thesis but they are often generated with the help of the boundary representation as in the recent work of Zhang et al. [ZWH12].

Due to the widely-used applications, surface modeling has been an active field of research over the last decades. We categorize existing modeling techniques into three classes: simplicial, parametric and implicit approaches, cf. [GVJ$^+$09]. All methods of these categories have advantages and disadvantages, therefore a best practice is application dependent. We summarize some significant works with an emphasis on parametric methods because the modeling technique we introduce in this thesis belongs to that class.

2.1.1 Simplicial surfaces

A mesh-based surface representation is defined by a collection of vertices and their topology or connectivity which induces facets or polygons, often triangles. The vertices and their connectivity are a piecewise linear approximation of the surface to be modeled. The extraction of a simplicial mesh from an unstructured point cloud is a delicate problem on its own. A short summary on existing techniques can be found in [GVJ$^+$09]. Nevertheless, a facet model offers limited accuracy, unless it uses a very dense point cloud which leads to a demanding data memory space, cf. [Ahn04]. For this reason, considerable research work deals with the question of mesh decimation, which compresses 3D data in order to allow an economic organization, storage and transmission, cf. [HG00]. Additionally, there is a need for constrained-based mesh editing. In general, one wants to locally deform a given mesh while satisfying certain constraints, see [DBD$^+$13]. Examples could be the planarity of the polygons in case they are not triangles or the preservation of the connectivity. Signal processing for meshes is one famous example for mesh deformation. Smoothing of discrete curvature information of a mesh is reached with the help of decreasing the discrete Laplacian while keeping the same number of vertices and the same topology. An introduction to some discrete fairing techniques can be found in [HG00].

Meshes are commonly used in computer graphics since they provide a flexible representation for objects of arbitrary topology. Nevertheless, smooth surfaces can only be modeled in a discrete sense, see [HG00], therefore the access to higher order information such as gradients or curvature need a thorough analysis. Meshes often serve as a source for the generation of parametric or implicit representations.

2.1.2 Parametric surfaces

Common parametric realizations of two-dimensional surfaces in $\mathbb{R}^3$ are functions defined on rectangular domains of $\mathbb{R}^2$. A widely used example are tensor product B-splines or their generalization NURBS (**N**on-**U**niform **R**ational **B**-**S**pline) patches. The latter play a key role in computer-aided design, manufacturing and engineering because of their usability and flexibility. As pointed out by Peters in [Pet95], they guarantee a certain smoothness, provide a compact and efficient representation as well as a geometrically intuitive tool for surface deformations. However, surfaces of arbitrary topology can not be parametrized on a single rectangular domain without singularities, [SXG$^+$09], unless each point of the underlying mesh is regular which means adjacent to exactly four quadrilaterals, cf. [Pet95]. This is due to the tensor product construction which induces a rectangular parameter domain, [Cas10]. To overcome this major drawback, one usually deals with local parametric patches which are trimmed and glued together. This glueing or rather stitching is expensive, prone to numerical error and in general it fails to provide the interior smoothness of the patches along the seams, cf. [DKT98].

Since classic tensor product splines lack local detail control, [DLP13], much research effort concentrates on local refinement methods for splines. Forsey and Bartles first introduced hierarchical overlays, [FB88, FB95]. The basic idea is to locally add finer B-splines to coarse ones. Further works extended this idea and replaced coarse B-splines by finer B-splines without creating the global impact induced by knot insertion, so-called hierarchical B-splines, and promoted the analysis, [Kra98, VGJS11]. T-splines were introduced to decrease degrees of freedom compared to hierarchical splines, [SZBN03]. They are defined on a quadrilateral mesh with T-junctions but elaborate attention must be spent on guaranteeing linear independence and other theoretical aspects of the underlying spline space, see [VGJS11]. To overcome some of these problems, Dokken et al. suggest the usage of LR-splines (**L**ocal **R**efinement), which result from a sequence of inserted line segments parallel to the knot lines. The bivariate case is discussed in [DLP13]. Although this approach guarantees linear independence, an analysis for higher dimensions is more sophisticated and the authors owe the reader a refinement strategy.

An elegant representation, especially for closed surfaces, which generalizes spline patches to arbitrary topology, are subdivision surfaces, [BMZB05]. Subdivision may generate surfaces recursively as the limit of a sequence of meshes. In each step a mesh is replaced by a denser one by computing new vertices with a function (*refinement rule*) of the old vertices, cf. [Cas10]. The benefit of this approach is the possibility to adapt the refinement rule in case the mesh contains irregular vertices. Since this approach only acts on a mesh, one might think that subdivision belongs into the category of the previous section. However, in case the refinement rule is derived from B-spline knot insertions, one has access to an analytic analysis of the limit surface, see [Cas10]. That is the reason why subdivision surfaces can rather be interpreted as 'spline surfaces with singularities', cf. [PR08], and are therefore parametric realizations. This denotation already indicates the price to pay for the topological freedom: irregular or extraordinary vertices prevent many simple and elegant refinement rules from generating C^k limit surfaces with $k \geq 2$ for arbitrary meshes without any flat points, cf. [PU98, KPR04, PR08, SXG$^+$09]. Nevertheless, due to its simplicity, subdivision has been used extensively within the animation industry since the late nineties, [DKT98].

Manifold-based constructions for smooth surface fitting have been a solution to overcome the seams caused by abutting surface patches. The basic idea is to separate the geometric and the topological problem: First, one determines a so called *proto-manifold* [SXG$^+$09] which defines the surface's topology by a *proto-atlas*. The latter does not provide the actual embedding but solves the segmentation problem of the underlying mesh and provides transition-functions which specify the overlap regions, cf. [GZ06]. Second, one obtains each embedding chart through a fitting procedure of the mesh part that lies in its domain. Finally, blending functions weight the influence of the charts respecting the overlap regions. For a nice overview and introduction to applications we refer to [GZ06]. Many works on manifold-based approaches usually lack an intricate mechanism to define the transition functions, unhandy many charts or a quadrilateral mesh, see [SXG$^+$09]. Manifold splines were introduced in [GHQ06] to overcome some of the mentioned problems: this construction uses affine transition functions and yields purely (rational) polynomial parametrization with the help of triangular B-splines. Nevertheless, unavoidable singularities of closed non-toroidal surfaces have to be cut out and replaced by hole-filling techniques. Later, T-splines and manifold splines were consolidated in [HWW+06] to obtain local detail control. The work on polycube splines [WHL+07] extends the manifold splines with a recipe for level-of-detail control and an easier chart construction. The method of Siqueira et al. avoids singularities, works with triangular B-splines, reduces implementation complexity but unfortunately does not generate polynomial surfaces, cf. [SXG+09].

Grimm et al. moved away from the idea of sticking to a mesh structure, which lacks a segmentation procedure as well as the definition of transition functions. Instead, they propose to use canonical surfaces of the appropriate topology (sphere, g-holed tori, $g > 0$) as global domains for the constructed surfaces, [GH03, GZ06, GJPH09]. This idea had been used before, e.g. in [WP97, RP99], but they treat the domain as a single entity with multiperiod functions. Contrary, Grimm et al. use fixed atlases for the canonical surfaces, bijectively map the target mesh to its topological equivalent and generate local parametrization while also interpreting the target surface as three-dimensional function on the canonical surface. They choose polynomial blending and embedding functions. Transition functions are induced by the fixed atlases and adaptivity is provided by iteratively adding charts. The first part of their approach is similar to the method presented in this thesis.

A method that completely avoids the detour via planar charts is presented in [CLCQ12]. The authors use so-called spherical Delaunay configuration B-splines as function space on the surface, an adaption to triangular B-splines that avoids partitioning discontinuity and singularities. Though the method provides adaptivity, effort is spent on generating a spherical Delaunay configuration.

An explicit homeomorphic mapping of a target surface onto a canonical surface is not a trivial task. Algorithms for generating spherical parametrization, which are especially interesting for texture mapping, have received much attention, a short overview is given in [HPS08]. First approaches were numerically unstable and could not guarantee bijective results for the discrete case. The authors of [GGS03] successfully generalize the method of barycentric coordinates to the sphere. This method guarantees bijectivity but also profits from the results of [SYGS05] for efficiently solving the resulting system. A multi-resolution technique is used in [PH03]. The approach simplifies the model via mesh-decimization, embeds it on the sphere and reinserts vertices iteratively. The latter is driven by minimizing the stretch metric of the parametrization. Other works strive for finding conformal mappings onto the sphere, e.g. [GWC+04]. Some analoga for non-spherical topology also apply a multiresolution technique by mapping decimized meshes of the same topology and iteratively refining the meshes, see e.g. [SAPM04]. Their algorithm does not require that the two meshes to be mapped contain the same number of vertices. By choosing some mesh on a torus one could obtain bijective mappings to a genus one model although the algorithm's execution time is unsatisfactory, see [SAPM04]. Other works even provide cross-parametrizations for models of different topology, e.g. [LYC+06, BPJ08]. These methods are naturally limited if meshes to be mapped differ too much in shape, cf. [HPS08]. Grimm et al. use an approach which first generates a greedy homotopy basis of $2g$ loops with a base point [EW05] and cuts the mesh open along the loops, cf. [GJPH09]. Then they correct the labeling of the loops until it can be embedded in a $4g$ sided polygon, cf. [GH03].

Parametrizations always involve the choice of parameters which largely affects the quality of the result, see [PL03]. Hoschek and Lasser address this problem and suggest an iterative parameter correction, see [HL93]. The main idea is to produce approximants whose error vectors are normal to the curve or surface to be modeled, cf. Section 8.1. Other works directly use a geometric fit by minimizing the square sum of the shortest distances, e.g. [PL03, Ahn04]. However, this non-linear procedure is numerically expensive and is therefore often simplified, for example through linear relaxations.

2.1.3 Implicit surfaces

A recurrent problem of the previous section is the emergence of singularities for non-toroidal surfaces. Implicit representations describe surfaces as well as planar curves as zero sets of a scalar-valued functions, overcoming topological bounds. Other significant advantages compared to parametrizations are the easy evaluation of intersections and the fact that an implicit representation already partitions the space into *inside* and *outside*. This means that implicit surfaces inherently model volume, therefore point classifications are trivial, cf. [BW97]. Moreover, the gradient of an implicit representation is always normal to the surface or curve. Early approaches are based on the idea of blending implicit primitives, for instance quadrics [GVJ+09]. Global implicit surface fitting techniques often try to reconstruct the signed distance function (see Subsection 7.3.2) with for example polynomials or radial basis functions, see [GVJ+09, OBA+03], although compactly supported functions are preferred from a computational point of view. In order to obtain local shape control, moving least squares is a promising method for implicit surfaces without sharp features, [Wen05, GVJ+09]. So-called partition of unity implicits use the same idea as manifold surfaces: local approximants are blended together by smooth weight functions which sum up to one. For instance, an adaptive technique which even handles sharp features is introduced in [OBA+03]. For a nice introduction to the advantages and usage of implicit surfaces in applications, we refer to [BW97]. The price to pay for the topological flexibility of implicit representations is an increased effort for a visualization of the modeled object. The usefulness of implicit representations is demonstrated by the so-called *level set method* pioneered by Osher and Sethian, see for example [Set96, OF02]. Due to the topological freedom, it addresses the modeling of moving interface problems. A comprehensive introduction can be found in [OF02].

2.2 Approximation of functions on manifolds

As already foreshadowed in the previous section, some surface modeling techniques can be easily adapted for general approximation of functions on manifolds. All approaches that create function bases over objects and interpret the target surface's coordinates as function values to be fitted are suitable. More precisely, the manifold-type constructions of the previous section have the conceptual advantage that the actual embedding step is independent of the dimension, cf. [DVJK08]. Considerably many approaches treat the case of the sphere or sphere-like surfaces [BLS05, CLCQ12, ANS96]. Another ansatz are spherical harmonics, see e.g. [Sch05, AH12] and Subsection 4.3.1. They form a complete and smooth function space on the sphere and are an inherent part of computer graphics and earth science. Spherical harmonics are eigenfunctions of the Laplace-Beltrami operator and there are attempts to adapt this approach to arbitrary manifolds [Lev06, Rus07]. However, the generalization is nontrivial and numerically expensive. Projected Powell-Sabin splines [DS07] have been also used for functions on two-manifolds of arbitrary topology but only generate C^1-approximants. All these approaches share the general idea of constructing a function space on the surface itself.

A completely different approach is proposed in [BOP92]. The authors interpolate scattered data with the help of so-called *fat surfaces*. Together with vertices of a triangulation, normal vectors at the vertices and estimated gradients, they extend the function to be interpolated to a neighborhood of the underlying surface and generate trivariate C^1-interpolants. Ruuth et al. introduce the closest point method for solving partial differential equations on surfaces, see [RM08], a fundamental enhancement to the idea of embedding a function on a surface into its ambient space. The method first extends the partial differential equations constantly into normal direction within a narrow band of the surface at hand. Secondly, it applies simple finite difference methods since surface gradients are replaced by standard gradients. Further research work concerning partial differential equations on surfaces followed this idea, for instance by the usage of radial basis functions [Pir12].

3 The Ambient B-Spline Method

We characterize our idea for the approximation of functions on submanifolds. The new approach enables us to easily generate C^k-approximants for arbitrary $k \in \mathbb{N}$. Additionally, the method inherently provides the possibility to generate C^k-surfaces. Some aspects are predicated on similar ideas of the work of the closest point method in [RM08] as well as the first step of the works of Grimm et al. [GH03, GZ06, GJPH09]. However, in contrast to many existing techniques, we neither derive complicated function spaces on the submanifolds individually nor face topological limitations although the approximation bases on a simple tensor product structure.

3.1 Basic concept

Here and below, let $\mathcal{M}$ denote a compact (boundaryless), embedded, $\bar{d}$-dimensional C^k-submanifold of $\mathbb{R}^d$ with $k \geq 2$. This means that for each point $y \in \mathcal{M}$ there exists a C^k-chart φ with $\varphi : \mathcal{M} \mapsto \mathbb{R}^d$ and an open neighborhood $y \in U \subset \mathbb{R}^d$, such that $\varphi(U \cap \mathcal{M})$ equals the intersection of some $\bar{d}$ - dimensional plane with $\varphi(U)$, see for example [Lee05]. In the following, we will always refer slovenly to (reference) *manifolds* or even to surfaces, when we actually mean submanifolds of the type just introduced. The word *reference* refers thereby to its role as preimage of the approximation. Throughout this thesis, we only consider codimension one, i.e. $\bar{d} = d - 1$. Experiments indicate that the ambient approximation method below is also applicable for manifolds of higher codimension, but a proof for the error analysis is ongoing research, cf. next section.

Let $g : \mathcal{M} \mapsto \mathbb{R}^{\bar{d}}$ be a function on $\mathcal{M}$, either given explicitly or via discrete data points. Our goal is to find a reasonable approximation without constructing complicated function spaces on the manifold itself. The basic idea of our approach is the transformation of the approximation on the manifold $\mathcal{M}$ to an approximation on a manifold's neighborhood $\Omega \subset \mathbb{R}^d$. Figure 3.1 illustrates the general modus operandi: first, the function g is extended to some neighborhood Ω in the ambient space of $\mathcal{M}$. Second, the extended function G is approximated by some approved technique for subdomains of $\mathbb{R}^d$. Finally, the restriction of the approximation of G to the manifold $\mathcal{M}$ provides the demanded approximation of g.

This procedure is fairly simple because approximation on subdomains of $\mathbb{R}^d$ is well understood and our approach does not rely on the given parametrization of the reference manifold. Additionally, the method is independent of the manifold's topology and therefore flexibly applicable in contrast to many approaches described in Chapter 2. Two ad-

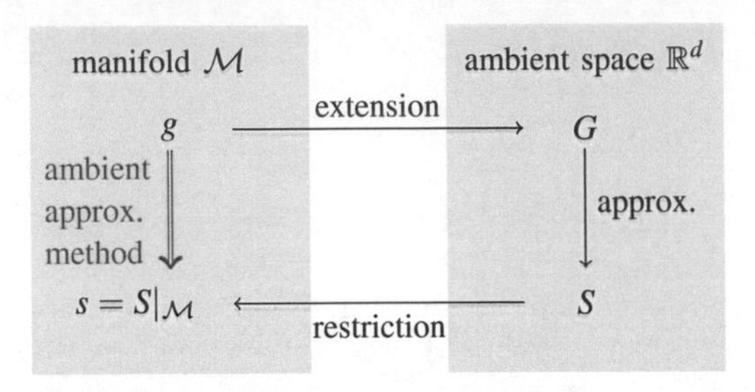

Figure 3.1: Flowchart of the ambient approximation method.

justing levers remain: We have to specify the construction of the extension and choose an approximation method. As the title of this thesis already suggests, for the latter we are going to use tensor product B-splines which are one of many possibilities of approximation methods based on a partition of the underlying domain. These piecewise polynomial functions with local supports provide good approximation properties, cf. Section 3.2, form a stable basis when used on certain domains of $\mathbb{R}^d$ and are easy to handle. A comprehensive work on meshless alternatives for scattered data approximation is provided in [Wen05]. However, the price to pay for the embedding in the ambient space is of course the increase in the number of degrees of freedom which roughly quantifies by a fixed factor n, where n refers to the B-spline order, see Section 4.1.

We address the details of the B-spline usage more elaborately in Chapters 4 and 5 and focus on the extension for the rest of this section.

It is clear that an extension of the function g on $\mathcal{M}$ into the ambient space is necessary, because approximation techniques in $\mathbb{R}^d$ of $(d-1)$-dimensional objects are unstable and may lead to numerically under-determined linear systems and therewith to undesired approximants. The task of the extension is therefore to secure the stability of the solved linear system, see Definition 4.3.

Concerning the choice of the neighborhood Ω, using tensor product B-splines for the approximation requires Ω to be composed of boxes defined within a narrow band near the manifold, as is similarly described in [RM08]. Such a narrow band is better known as *tubular neighborhood* of a manifold. This neighborhood fulfills the so-called *nearest* or *closest point property*, see [Foo84]. A neighborhood Ω of $\mathcal{M}$ has this property, if for all $x \in \Omega$ there exists a unique point $\mathrm{clp}(x) \in \mathcal{M}$ with minimal distance: $\min_{y \in \mathcal{M}} \|x - y\| = \|x - \mathrm{clp}(x)\|$. Thereby, the function clp projects a point x along the corresponding normal vector of $\mathrm{clp}(x)$ onto the manifold. Amongst others, the existence of such a neighborhood is proven in [Foo84]:

Lemma 3.1. *Let* $\mathcal{M} \subset \mathbb{R}^d$ *be a manifold as defined above. Then* $\mathcal{M}$ *has a neighborhood* Ω *with the closest point property, and the projection map* $\mathrm{clp} : \Omega \to \mathcal{M}$ *is* C^{k-1}.

With the help of this theoretical result, extending the function g constantly in normal direction within a tubular neighborhood is a canonical choice and has been considered before, see [BOP92, RM08]. For instance, in this setting the Laplace-Beltrami operator of the reference manifold is consistent with the Laplace operator in the ambient space, [RM08]. For practical reasons, of course one needs to quantify the narrowness or determine a globally valid radius r of such a neighborhood, respectively. The so-called *medial axis* of a manifold is sort of the counterpart, as it is the set of all points, which have more than one closest point on the manifold and therefore cannot belong to a tubular neighborhood.

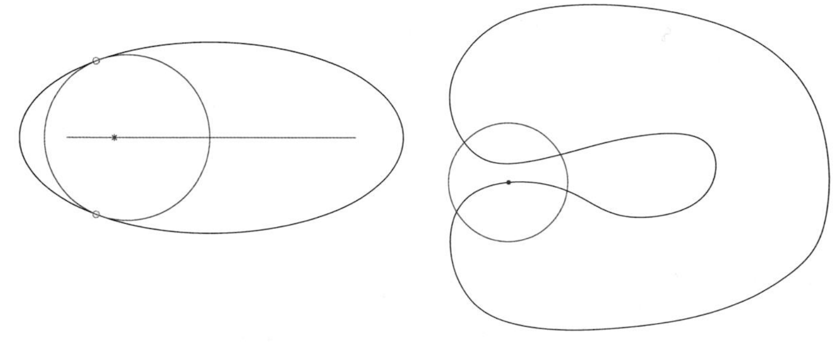

(a) The red line – marks the medial axis of the ellipse. The two points circled magenta ∘ are both the closest point of the blue point $*$. The minimal distance of the medial axis to the ellipse equals $\frac{1}{|\kappa_{\max}|}$.

(b) Example of a closed curve in $\mathbb{R}^2$. The circle around the blue point • indicates a neighborhood if only curvature information is considered. Since two connected components come close in the Euclidean sense, it is not enough to only consider the maximal absolute curvature to obtain a valid tubular neighborhood.

Figure 3.2

The authors of [KS08] provide an analytic derivation for a valid radius of C^2-surfaces. The proof bases on the famous theorem from convex analysis which guarantees existence and uniqueness of the closest point projection. If we envision some simple examples, such as an ellipse in $\mathbb{R}^2$, it is clear that a valid radius is locally bounded by the inverse of the maximal absolute curvature (principal curvature for $d > 2$) of the reference manifold:

$r < \frac{1}{\kappa_{\max}}$, see Figure 3.2 **(a)**. Considering a slightly more complex example, we realize that a neighborhood of a point bounded by the locally maximal curvature may contain two connected components of a curve, see Figure 3.2 **(b)**. In this case, a valid radius may be significantly smaller than $\frac{1}{\kappa_{\max}}$. This is the reason why we suggest to choose reference manifolds whose tubular neighborhoods are globally determined by the curvature information when we want to model surfaces.

Instead of extending the function into normal direction, we could also choose any C^{k-1} vector field V on the manifold $\mathcal{M}$, which is not tangent. Then one has to verify that the function $\eta : \mathcal{M} \times [-r,r] \to \Omega_r$, $\eta(x) := x + tV(x)$ is a diffeomorphism if r is small enough. The proof for the special case that V is the normal vector field can be adapted, even though $|t|$ then does not comply to the shortest distance between $x + tV(x)$ and the manifold $\mathcal{M}$ in the Euclidean sense. Since the normal vector field is easily at hand, we do not investigate on other options any further but point out that the method provides a certain amount of flexibility with regard to the extension direction.

Finally, for the extended function G holds

$$G : \Omega \mapsto \mathbb{R}^{\tilde{d}},$$
$$G(x) := g(\text{clp}(x)).$$

For greater clarity for upcoming sections, we introduce an extension as well as a restriction operator $\mathcal{E}$ and $\mathcal{R}$ with $\mathcal{E}(g) := G$ and $\mathcal{R}(G) := g$.

3.2 Approximation power

Before elaborating the details of the approximation process, we need to assure that the construction of the previous section provides satisfactory approximation power. The rather technical proof for codimension one is prepared in current research work of Odathuparambil, Prasiswa and Reif. A proof for higher codimension cannot be established by the same arguements and is still an open problem. We only mention the main result which guarantees the full approximation power for the ambient B-spline method.

We assume the reader is familiar with classic Sobolev spaces and its corresponding norms and seminorms and refer to the book [Ada75] for a comprehensive introduction.

Although the ambient method does not depend on the choice of the charts of the underlying reference manifold $\mathcal{M}$, for an error analysis we need a concrete atlas. So let $(U_i, \varphi_i)_{i \in \mathcal{I}}$ be a finite set of charts covering $\mathcal{M}$. Before providing the error estimate for the ambient B-spline approximation, we need to define Sobolev spaces with their corresponding norms on manifolds. These definitions have been used before in different contexts and can be found for example in Subsection 3.6.1 of the book [Tri78]:

Definition 3.2. *For $1 \leq p \leq \infty$, the Sobolev-norm of order n of a function g on a C^k-manifold $\mathcal{M}$ is given by*

$$\|g\|_{W_p^n(\mathcal{M})} := \left(\sum_{i \in \mathcal{I}} \|g \circ \varphi_i^{-1}\|_{W_p^n(U_i)}^p \right)^{\frac{1}{p}}, \quad p < \infty$$

and

$$\|g\|_{W_\infty^n(\mathcal{M})} := \max_{i \in \mathcal{I}} \|g \circ \varphi_i^{-1}\|_{W_\infty^n(U_i)}.$$

Then

$$W_p^n(\mathcal{M}) := \left\{ g : \mathcal{M} \to \mathbb{R} : \|g\|_{W_p^n(\mathcal{M})} < \infty \right\}$$

defines a Sobolev space on $\mathcal{M}$.

The parameter n denotes the maximal order of partial derivatives. In the following, n also refers to the order of the underlying B-spline approximation, cf. Chapter 4, because the B-spline approximation error is usually bounded by the corresponding Sobolev seminorm.

In case $g : \mathcal{M} \mapsto \mathbb{R}^{\tilde{d}}$ is a vector-valued function, i.e. $\tilde{d} > 1$, we may treat each component independently. One can construct a spline approximation S of a function G on a box $\omega \subset \Omega \subset \mathbb{R}^d$ with *locally optimal* error behavior, see e.g. [Sch81]:

$$\|G - S\|_{W_p^m(\omega)} \leq ch^{n-m} |G|_{W_p^n(\omega)}, \tag{3.1}$$

where h refers to the used grid size, which is chosen equally for all coordinate directions throughout this thesis. The term $|G|_{W_p^n(\omega)}$ refers to the Sobolev seminorm which means that only the derivatives of order n influence the error estimate. With these consideration, we may state the theorem that guarantees the desired approximation power of the ambient B-spline method:

Theorem 3.3 (Odathuparambil, Prasiswa, Reif)**.** *Let $1 \leq p \leq \infty$, $g \in W_p^n(\mathcal{M})$ and $G = \mathcal{E}(g)$ as defined in the previous section. Let S be a locally optimal spline approximation for G and $s := \mathcal{R}(S)$. If h denotes the grid size, it holds*

$$\|g - s\|_{W_p^m(\mathcal{M})} \leq ch^{n-m} \|g\|_{W_p^n(\mathcal{M})}$$

for some constant c and $m < n$.

Whereas standard tensor product B-splines are well suited for approximating functions defined on boxes that only contain full grid cells, concerning the approximation on general domains $\Omega \subset \mathbb{R}^d$ they usually lack stability and may forfeit approximation power. The problems arise because tensor product grids are not adaptable to the treated domains in contrast to the univariate case. Various works deal with the elimination of the stability problem. For instance, extended B-splines, e.g. [HRW01, Höl03, Pra09, DPR13] are a possibility to guarantee stability but unfortunately the constructed spaces are not nested in contrast to the standard tensor product spline space. Another idea are normalized B-splines, [Mößo6], but sometimes isolated B-splines still remain unstable. The procedure of skipping the B-splines concerned may lead to a loss of approximation power in L_p spaces for arbitrary space dimensions. An approach which keeps the nestedness and generates stable basis elements for the approximation on Lipschitz domains are the so-called condensed B-splines constructed in [Sis11] but are only applicable in $\mathbb{R}^2$.

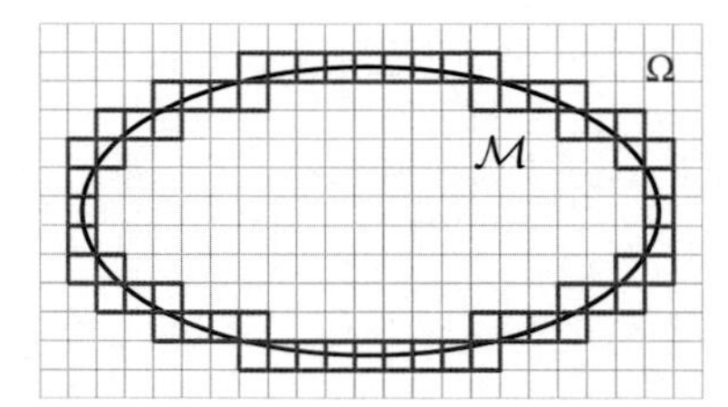

Figure 3.3: The neighborhood $\Omega := \bigcup_i \square_i$ for the extension only consists of those cells of the corresponding grid of the B-spline basis which intersect the manifold $\mathcal{M}$, cf. Section 4.2.

Due to the simple structure of standard tensor product B-splines, our course of action adapts the approximation domain that not only stability without any modifications, cf. Chapter 4, but also full approximation power is guaranteed. We establish such a domain as a finite union of full grid cells, see Figure 3.3. On each such cell, local optimal approximation power holds. We can therefore construct quasi-interpolants, see Section 4.1, and apply the famous *Bramble-Hilbert Lemma*, cf. its introduction in [BH70]. The latter provides an error estimate for the approximation of a function by a polynomial of order n in Sobolev spaces for domains that fulfill the so-called strong cone condition. When treating approximations on more general domains one needs less restrictive results as elaborated in

[Rei12]. Additionally, we point out that the error estimate (3.1) depends on the ratio of the coordinate-wise grid sizes in the context of anisotropic Sobolev spaces, i.e., when different orders for each coordinate direction are considered. The proof of the Bramble-Hilbert Lemma on general domains gets difficult for anisotropic Sobolev spaces. We therefore only treat the isotropic case which means the same order for all coordinate directions and we justify this choice in the upcoming chapter. For a detailed discussion on anisotropic Sobolev spaces on more general domains we refer the reader to [Sis11].

4 Realization of the Ambient B-Spline Method

In this chapter, we first address the function spaces we use for the approximation in ambient space. Second, we describe the procedure of the ambient B-spline method in detail and illustrate the main stages by a simple, two-dimensional example. The last section presents the results of three examples in $\mathbb{R}^3$. We discuss their promising outcome as well as arising difficulties and suggest an improvement for the approximation.

4.1 Preliminaries

The famous *Weierstrass Approximation Theorem* states that continuous functions on compact sets can be uniformly approximated by polynomials, see e.g. [Sch81]. Nevertheless, possible high polynomial degrees and the instability of the monomial basis degrade the theorem to be more of theoretical interest rather than being useful in practice. A remedy for the mentioned difficulties is the usage of *spline functions*. Splines are piecewise polynomial functions with local supports. They were introduced to the approximation community already in the forties, cf. [Sch46]. Also due to their benefits for various industries, see Section 2.1, extensive research focused on spline theory in the following decades.

We briefly introduce the B-spline basis and some theoretical tools for its analysis in the following two subsections. B-splines were especially promoted by Carl de Boor, whose monograph [dB78] is one of the standard references and to which we refer the reader for a comprehensive introduction.

4.1.1 B-splines

A univariate *spline* s of order n is a real function which consists of piecewise polynomials of order n, i.e., of maximal degree $n-1$. Its name has been first coined by Schoenberg, cf. [Sch46]. The individual polynomial pieces are bordered by *knots* $\tau_k \in \mathbb{R}$, i.e. if $\tau_k < \tau_{k+1}$, the function s equals a polynomial on the half open interval $[\tau_k, \tau_{k+1})$. For convenience, we assume the so-called *knot sequence* $\mathcal{T} = \{\tau_k\}_{k\in\mathbb{Z}}$ to be bi-infinite and monotonic increasing, i.e. $\tau_k \leq \tau_{k+1}$ for all $k \in \mathbb{Z}$. Additionally, we take the knot sequence $\mathcal{T}$ to be *non-degenerate*, that means $\tau_k < \tau_{k+n}$ for all $k \in \mathbb{Z}$. The spline s is at least $n-1-i$-times differentiable at a knot τ_k with knot multiplicity i. All splines with knot sequence $\mathcal{T}$ form the spline space $\mathcal{S}^n(\mathcal{T})$. There are various bases for the spline space $\mathcal{S}^n(\mathcal{T})$, but we only use and shortly introduce the *B-spline* basis.

B-spline basis functions can be introduced via recursion [dB78] or with the help of the truncated power basis [Sch81]. We proceed analogously to [Möß06] and first introduce the following set of functions:

Defintion and Lemma 4.1. *Let $n \in \mathbb{N}$ and $\mathcal{T}$ be a non-degenerate knot-sequence as defined above. We define the associated* ψ-functions *as polynomials of order n with $n-1$ real roots:*

$$\psi_k^n(x) := \prod_{i=1}^{n-1}(\tau_{k+i} - x), \quad k \in \mathbb{N}.$$

If $\tau_k < \tau_{k+1}$, the functions

$$\psi_{k-n+1}^n, \psi_{k-n+2}^n, \ldots, \psi_k^n$$

are linearly independent.

The proof of the latter verifies that the corresponding n ψ-functions may generate the monomials, cf. e.g. [Möß06]. This definition enables us to formulate *Marsden's identity*:

Defintion and Theorem 4.2. *Let $\mathcal{T}$ be a knot sequence as defined above of a spline space of order n. Then there exist functions B_k^n which are uniquely determined by their support,*

$$\operatorname{supp}(B_k^n) \subseteq [\tau_k, \tau_{k+n}],$$

and the equation

$$(t-x)^{n-1} = \sum_k B_k^n(t)\psi_k^n(x), \quad t \in \mathbb{R}, x \in \mathbb{R}.$$

The functions B_k^n are called B-splines and build a basis of the spline space $\mathcal{S}^n(\mathcal{T})$.

A spline $s : \mathbb{R} \mapsto \mathbb{R}^d$ is a linear combination of B-splines,

$$s(t) = \sum_k b_k^n B_k^n(t),$$

where the spline coefficients b_k^n are also called *control points*. For example, $b_k^n \in \mathbb{R}^2$ for representing curves in $\mathbb{R}^2$ or $b_k^n \in \mathbb{R}^3$ for curves in $\mathbb{R}^3$. The B-splines form a non-negative partition of unity and have compact support by definition. One can show the local and the global basis property. B-splines of order n can be expanded from B-splines of order $n-1$ with the help of a recursion, which is the key to the algorithm for B-spline evaluation. Differentiation and integration can be expressed as an operation on the control points. Together with the stability property, B-splines are well suited for the implementation of spline algorithms. One generally has a certain amount of flexibility concerning the choice of the knot sequence $\mathcal{T}$. However, we only consider uniformly spaced knot sequences in this thesis and therefore introduce the special knot sequence $\mathcal{T} = \mathbb{Z}$. The corresponding *cardinal* B-splines are illustrated in Figure 4.1. We notice that cardinal B-splines are integer shifts of each other. For upcoming definitions, we refer to $\mathbf{B}_o^n$ as the univariate, cardinal B-spline of order n with support on the closed interval $[0,n]$.

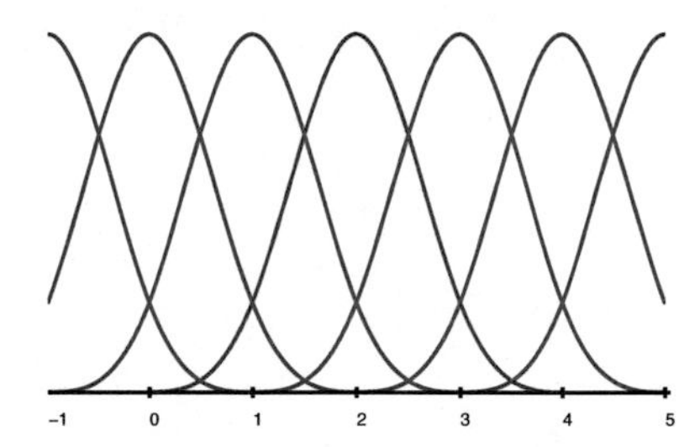

Figure 4.1: Cardinal B-splines ($n = 4$) are integer shifts of each other. $\mathbf{B}_o^n$ is marked in red —.

A straightforward generalization to multivariate splines on $\mathbb{R}^d$, $d > 1$, is their definition via tensor products, cf. [dB78, Möß06]. Using d non-degenerate knot sequences $\left[\mathcal{T}^1, \dots, \mathcal{T}^d\right]$, we obtain with

$$\mathcal{S}^n(\mathcal{T}_d) := \bigotimes_{\iota=1}^{d} \mathcal{S}^{n_\iota}(\mathcal{T}^\iota), \quad \mathcal{T}_d = \prod_{\iota=1}^{d} \mathcal{T}^\iota$$

the tensor product spline space of order $n = (n_1, \dots, n_d)$, where each entry of n then refers to the order of the corresponding component. Since $\mathcal{T}_d$ is the Cartesian product of the univariate knot sequences, it can be interpreted as d-dimensional grid. The tensor product B-splines

$$B_k^n(x) := B_{k_1}^{n_1}(x_1) \cdots B_{k_d}^{n_d}(x_d), \quad k \in \mathbb{Z}^d, \quad x \in \mathbb{R}^d,$$

form a basis for $\mathcal{S}^n(\mathcal{T}_d)$. All of the above mentioned properties persist and spline evaluation or differentiation can be realized by a d-fold application of the univariate algorithms.

4.1.2 Dual functionals and quasi-interpolants

As already stated in [Möß06], it is important to analyze the sensitivity of a spline, i.e. the impact of changes of the coefficients b_k^n on the spline itself. If one can derive uniform bounds of these deviations, the basis is called *stable* or *well conditioned.* Whereas the monomials are an example for a badly conditioned basis and therefore not suitable for numerical computations on large intervals, stability of the B-splines is proven in [dB78]. For completeness, we employ the definition given in [Möß06]:

Definition 4.3. *Let $\boldsymbol{B} = (B_k)_{k\in\mathbb{Z}^d}$ be a basis of a normed vector space $(V, \|\cdot\|)$ and $|||\cdot|||$ be a norm on the coefficients $\boldsymbol{b} = (b_k)_{k\in\mathbb{Z}^d}$. If*

$$c := \inf_{\boldsymbol{b}\neq 0} \frac{\|\sum_{k\in\mathbb{Z}^d} b_k B_k\|}{|||\boldsymbol{b}|||} \quad \textit{and} \quad C := \sup_{\boldsymbol{b}\neq 0} \frac{\|\sum_{k\in\mathbb{Z}^d} b_k B_k\|}{|||\boldsymbol{b}|||}$$

are finite, then it holds

$$c|||\boldsymbol{b}||| \leq \left\| \sum_{k\in\mathbb{Z}^d} b_k B_k \right\| \leq C|||\boldsymbol{b}||| \quad \textit{for all } (b_k)_{k\in\mathbb{Z}^d}.$$

The quotient $cond(\boldsymbol{B}) := \frac{C}{c}$ is the condition *of the basis $\boldsymbol{B}$.*

The smaller the condition of a basis the better is its numerical behavior. The main ingredients of the proof for the stability of the B-splines, cf. for example [dB78, Möß06], are the linear *de Boor-Fix functionals* $\lambda := (\lambda_j)_{j\in\mathbb{Z}^d}$, which were introduced in [dBF73]. These functionals are *dual* to the B-splines, i.e., it holds

$$\lambda_j(B_k^n) = \delta_{jk},$$

where

$$\delta_{jk} = \begin{cases} 1, & j = k, \\ 0, & \text{otherwise.} \end{cases}$$

This leads to the possibility to directly compute the coefficient of a spline:

$$\lambda_j(s) = \lambda_j(\sum_k b_k^n B_k^n) = \sum_k b_k^n \lambda_j(B_k^n) = b_j^n.$$

It turns out that the *de Boor-Fix functionals* are very useful for analytical studies:

Defintion and Theorem 4.4. *Let $\mathcal{T}_d$ be a valid knot sequence as introduced above for a spline space of order $n = (n_1, \ldots, n_d)$. For $k \in \mathbb{Z}^d$ and $\tau \in \left[\tau_{k_1}, \tau_{k_1+n_1}\right) \times \cdots \times \left[\tau_{k_d}, \tau_{k_d+n_d}\right)$ we define*

$$\begin{aligned}\lambda_k(s) &:= \sum_{j_1<n_1} \cdots \sum_{j_d<n_d} \frac{(-1)^{n_1-1-j_1}\psi_{k_1}^{(n_1-1-j_1)}(\tau_1)}{(n_1-1)!} \cdots \frac{(-1)^{n_d-1-j_d}\psi_{k_d}^{(n_d-1-j_d)}(\tau_d)}{(n_d-1)!} s^{(j)}(\tau) \\ &=: \frac{1}{(n-1)!} \sum_{j<n} (-1)^{|n-1-j|} \psi_k^{(n-1-j)}(\tau) s^{(j)}(\tau) \quad j \in \mathbb{Z}^d.\end{aligned}$$

These functionals are dual *to the B-splines B_k^n. We refer to λ as the* de Boor-Fix functionals.

The second row of the definition is simply an abbreviation. Verifying the dual basis property requires the multivariate analogon of Marsden's identity and utilizes the fact that the B-splines form a local basis. These functionals also offer the possibility to derive the affine invariance under reparametrizations. A proof for the univariate case can be found in [Möß06].

We already indicated the good approximation power of splines, cf. Section 3.2. The proof bases on the usage of *quasi-interpolants*, which provide not only a nice tool for the analysis but also for a concrete implementation of local methods to increase numerical performance, cf. [LLM00].

Definition 4.5. *Let $(Q_k)_{k\in\mathbb{Z}^d}$ be a family of linear functionals. The operator $Q : C(\mathbb{R}^d) \mapsto \mathcal{S}^n(\mathcal{T}_d)$ is a* quasi-interpolant *of order m if the following three properties are satisfied:*

- *$Qg := \sum_k (Q_k g) B_k^n$, for $g \in C(\mathbb{R}^d)$ and the functionals Q_k only depend on the support of the corresponding B-spline, i.e. $Q_k : C(D_k) \mapsto \mathbb{R}$, where $D_k \subseteq \operatorname{supp}(B_k^n)$,*
- *the functionals Q_k are uniformly bounded, i.e. $\sup_{k\in\mathbb{Z}^d} |Q_k| \leq C$ for some constant C,*
- *polynomials of order m are reproduced, i.e. $Q\pi = \pi \; \forall \pi \in \Pi_m$.*

As is already pointed out in [dB78], the operator introduced with the help of the de Boor-Fix functionals fulfills the second property only for certain norms and if the treated function g is smooth enough. In case g does not comply to the requested smoothness, one can first construct an appropriate approximation g^* of g, for instance by polynomials, and then apply the quasi-interpolant to g^*, cf. [Pra09]. We briefly verify why λ may fulfill the three conditions of a quasi-interpolant. The first one is trivial since for $k \in \mathbb{Z}^d$ the choice of $\tau \in \left[\tau_{k_1}, \tau_{k_1+n_1}\right) \times \cdots \times \left[\tau_{k_d}, \tau_{k_d+n_d}\right)$ guarantees the locality. The third condition directly follows from the fact that the de Boor-Fix functional is a projector onto the spline space which reproduces polynomials. The second condition needs a little more effort. For $g \in C^{n-1}(\mathbb{R}^d)$, we obtain

$$|Q_k g| = |\lambda_k g| \leq \frac{1}{(n_1-1)!} \cdots \frac{1}{(n_d-1)!} \sum_{j<n} |\psi_{k_1}^{(n_1-1-j_1)}(\tau_1)| \cdots |\psi_{k_d}^{(n_d-1-j_d)}(\tau_d)| |g^{(j)}(\tau)|$$

$$\leq \underbrace{\frac{1}{(n_1-1)!} \cdots \frac{1}{(n_d-1)!} \max_{\iota=1}^{d} \max_{j_\iota=1}^{n_\iota-1} |\psi_{k_\iota}^{(n_\iota-1-j_\iota)}(\tau_\iota)|}_{(1)} \quad \underbrace{\sum_{i<n} |g^{(i)}(\tau)|}_{(2)}.$$

Expression (1) demands a detailed analysis of the univariate polynomials of Marsden's identity and its derivatives. Usually one sets the point τ to be in the largest cell of the underlying B-spline support. Since we deal with uniform grids, we randomly choose a cell $c_m := \left[\tau_{m_1}, \tau_{m_1+1}\right) \times \cdots \times \left[\tau_{m_d}, \tau_{m_d+1}\right)$. With the help of the affine invariance of the dual functionals, one chooses a skilled reparametrization ϕ and gets $\tilde{g} = g \circ \phi$. With this, one can derive a constant as upper bound which only depends on the order n, see [Möß06]. Of course the reparametrization also transforms Expression (2). We figure that each summand of (2) can be bounded by $\|\tilde{g}\|_{C^{n-1}(\tilde{c})}$, where $\tilde{c}$ refers to the domain on which $\tilde{g}$ is evaluated. Finally, we obtain the estimate

$$|Q_k g| = |\lambda_k g| \leq C(n) \|\tilde{g}\|_{C^{n-1}(\tilde{c})},$$

using that we deal with n^d summands. The last inequality approves the boundedness of the functionals.

4.2 Implementation

In this section we introduce in detail the proceeding of the algorithmic realization of the ambient B-spline method, implemented and tested in MATLAB©. Whereas the description is preferably general concerning space dimension or the given reference manifold, we illustrate the single steps by a simple example in $\mathbb{R}^2$ before studying more complex ones in Section 4.3.

We start by introducing uniform, orthogonal grids which we not only use as knot sequences for the determination of the B-splines but also for describing the domains on which we approximate. The following definition is based on ideas in [CCG+06].

Definition 4.6. *Let $\mathcal{G}_d$ be a disjoint partition of $\mathbb{R}^d$ into congruent d-dimensional boxes (cells), such that the origin of $\mathbb{R}^d$ is the lowest corner of some cell. $h \in \mathbb{R}_{>0}$ denotes the isotropic grid size of $\mathcal{G}_d$ and a cell $c_k \in \mathcal{G}_d$, $k \in \mathbb{Z}^d$ is defined as the half open box*

$$c_k := [k_1 h, (k_1+1)h) \times \cdots \times [k_d h, (k_d+1)h).$$

We associate to $\mathcal{G}_d$ the uniform, multivariate knot sequence

$$\mathcal{T}_d := h\mathbb{Z}^d.$$

We define the multivariate, uniform B-splines B_k^n of order $n = (n_1, \dots, n_d)$ with knot sequence $\mathcal{T}_d$ and index $k = (k_1, \dots, k_d) \in \mathbb{Z}^d$ by

$$B_k^n(x) := \mathbf{B}_o^{n_1}\left(\frac{x_1}{h} - k_1\right) \cdot \ldots \cdot \mathbf{B}_o^{n_d}\left(\frac{x_d}{h} - k_d\right),$$

where $\mathbf{B}_o^m$ is the univariate cardinal B-spline of order m, cf. Section 4.1. Consequently, its support is the d-dimensional closed box

$$\operatorname{supp}(B_k^n) = [k_1 h, (k_1 + n_1)h] \times \cdots \times [k_d h, (k_d + n_d)h].$$

For technical reasons, we want to assign a point $x \in \mathbb{R}^d$ uniquely to all relevant B-splines and therefore introduce the half open box σ_k^n of a B-spline B_k^n by

$$\sigma_k^n := [k_1 h, (k_1 + n_1)h) \times \cdots \times [k_d h, (k_d + n_d)h). \tag{4.1}$$

Definition (4.1) implies that we identify a B-spline B_k^n in $\mathbb{R}^d$ with its lower left corner of the corresponding half open box σ_k^n just as we do with the cells.
From now on we only consider isotropic spline spaces, thus it holds $n_1 = \cdots = n_d$. The parameter n can therefore be set as a natural number, i.e. $n \in \mathbb{N}$, and defines the B-spline order for all coordinate directions. Moreover, we neglect the index n and write B_k and σ_k instead of B_k^n and σ_k^n to improve readability. This procedure is justifiable because the reference manifolds only provide topological rather than geometrical information. An in-advance identification of a distinguished coordinate direction which should be treated by higher B-spline orders is therefore not possible.

With this preliminary work we are able to elaborate on the implementation of our method:

Input
At the beginning, a (reasonable) reference manifold $\mathcal{M}$ as defined in Section 3.1 and some function on it is given, see Figure 4.2. The function is either explicitly available or given via discrete data with an optional connectivity, such as a triangulation in case $d = 3$.

Grid size h
A valid grid size h has to guarantee that the emerging approximation domain as subset of $\mathbb{R}^d$ lies within a tubular neighborhood of the manifold $\mathcal{M}$, see Section

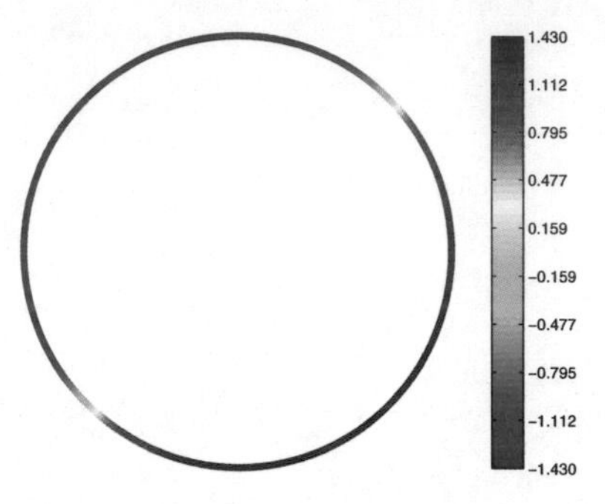

Figure 4.2: We consider a unit circle in $\mathbb{R}^2$ as a very simple manifold and the function $g(x,y) = \arctan(5(x-y))$ defined on it.

3.1. Excluding that two connected components of the reference manifold are close, we need to guarantee that the closest point projection of all points within the domain is well defined, $\|\operatorname{clp}(y) - y\| < \frac{1}{\kappa_{\max}}$ for all $y \in \Omega$, where $\kappa_{\max}$ is either the maximal absolute value of the principal curvatures or, in case we treat curves, the maximal absolute curvature.

Approximation domain Ω

Together with a chosen grid size h, we define the domain Ω as the set of all cells contained in $\mathcal{G}_d$ which have a non empty intersection with the manifold $\mathcal{M}$:

$$\Omega := \{c_k \in \mathcal{G}_d \,|\, c_k \cap \mathcal{M} \neq \emptyset,\ k \in \mathbb{Z}\}.$$

We call all cells contained in Ω *active cells*, see Figure 4.4.

Approximation order n

The choice of the order n for the B-spline approximation depends on the application and is influenced by the desired differentiability of the approximant and the computing time. For instance, the surface approximations of the Stanford Bunny and cow model in Section 4.3 are computed with $n = 4$ which leads to C^2-approximants, i.e., continuous curvature information.

Data sites

We fill each active cell with n^d uniformly distributed data sites, cf. Figure 4.5, in such a way that the minimal distance of two points along the coordinate axes equals

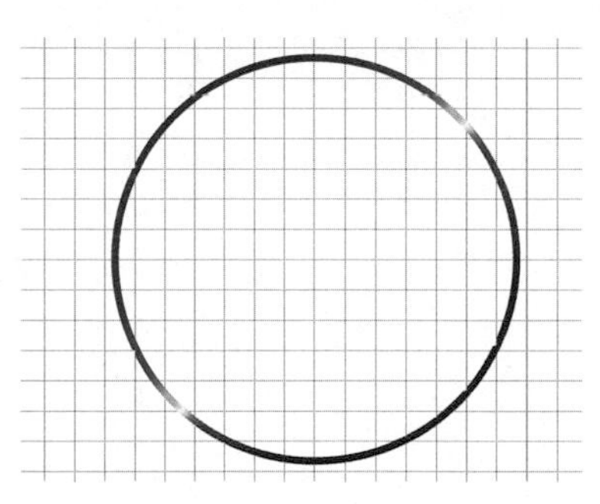

Figure 4.3: The chosen grid size for the example is $h = 0.15$.

$\frac{h}{n}$ and the minimal distance of some points to the boundary of the cell equals $\frac{h}{2n}$. With this alignment, the minimal distance of an arbitrary point to all its neighbors is always $\frac{h}{n}$. Distributing the data sites uniformly avoids problems when dealing with discrete least square problems, see the section below about the least square approximation and cf. Figure 4.6. We denominate the union of all data sites by $\mathcal{P}$.

Closest point projection

The determination of the closest points for all data sites on a given manifold is in general numerically expensive, unless the signed distance function or closest point projection is explicitly given, cf. Section 3.1. As already pointed out in Section 2.1, geometric fitting techniques use the latter to obtain parametrizations with preferably orthogonal error vectors.

There are a number of numerical approaches for the determination of the closest point. A short overview is given in [HW05], where the authors also suggest a second order algorithm. The usual procedure starts with an initial guess for a closest point on the manifold. We obtain our initial guess by a so-called zero order algorithm. That means we first search for the closest point within a discrete subset of the manifold. Since we subsequently apply a Newton-type iteration, we have to be aware of the sensitivity of initial values, cf. [HW05] and the references therein. We therefore need to assure an adequate (i.e. manifold dependent) density of the sampling in order to benefit from local convergence properties. The fundamental idea of a Newton-iteration is to solve a nonlinear system $A(x) = 0$, where the function $A : \mathbb{R}^d \to \mathbb{R}^d$ is C^1 and its derivative additionally fulfills a Lipschitz condition,

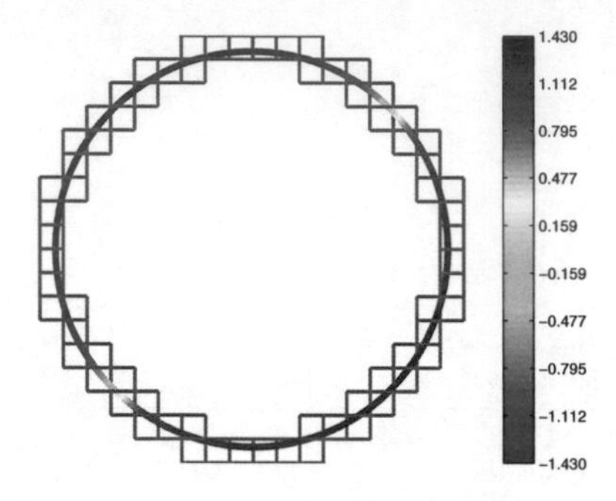

Figure 4.4: All active cells of the unit circle are framed blue —.

through a local linearization, i.e. $A(x^k) \approx A(x) + A'(x^k)(x - x^k)$ and x^k being an approximation. The update formula

$$x^{k+1} = x^k - (A'(x^k))^{-1} A(x^k)$$

is repeated until a sufficiently good approximation for the root is reached in case $A'(x^k)$ is regular. An analysis reveals the local quadratic convergence of the Newton approach, cf. [Höl03, Kan07]. We usually do not explicitly evaluate the inverse matrices but rather solve a system of linear equations to obtain a correction vector for updating the approximation x^k. The evaluation of the derivative in each iteration may still be numerically expensive. To overcome this drawback, Broyden suggested a rank one update in [Bro65], which means that the Jacobian is only computed once at the beginning and then successively updated. An elaborate introduction to this quasi-Newton method can be found in [Kan07]. Broyden's suggestion results in a diminishing of the convergence rate, which is only superlinear but may compensate for this by cheaper operations for each iteration. The nonlinear equation we have to solve in order to obtain an approximation for the closest point X^* of a point X in the ambient space of the reference manifold is

$$\left(\varphi'\right)^{-1}(x^*)\left(\varphi^{-1}(x^*) - X\right) = 0,$$

where φ is a chart of the atlas of $\mathcal{M}$ that parametrizes the corresponding region and x^* is the preimage of X^* in terms of φ^{-1}. With Broyden's method, we may even avoid to compute any second derivative of the charts by applying finite differences as initial guess. The numerical treatment of the closest point projection

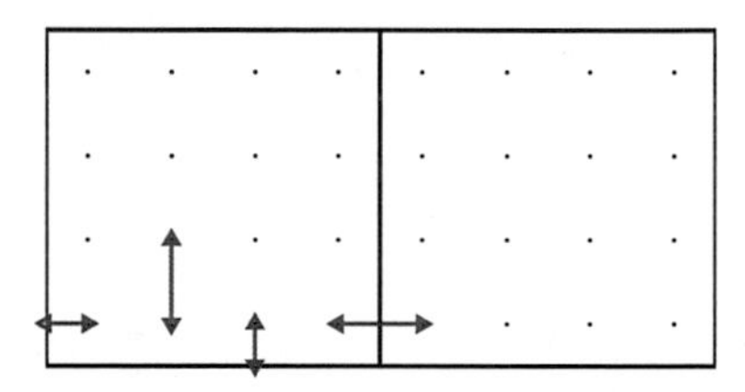

Figure 4.5: Two active neighbor cells in $\mathbb{R}^2$ with data cites $\mathcal{P}$ for $n = 4$. The blue double-headed arrows $\longleftrightarrow$ mark the distance $\frac{h}{n}$ and the red double-headed arrows $\leftrightarrow$ the distance $\frac{h}{2n}$.

is the only step of the ambient B-spline method which requires a parametrization of the reference manifold. The cases of the circle or the sphere, respectively, and the embedded torus are much simpler because their closest point projections can be derived analytically, cf. [RM08].

Assignment of function values to the projected data sites

In case the function g to be approximated is explicitly given, we assign to each $p \in \mathcal{P}$ the function value of the closest points on the manifold $G(p) = g(\mathrm{clp}_{\mathcal{M}}(p))$, see Figure 4.7. In case g is only given through discrete data, we need to approximate the function value using the underlying connectivity of the data, if available. For instance, for the examples of the Stanford Bunny and the cow model in Section 4.3 we use the given triangulation and obtain a function value for $p \in \mathcal{P}$ via linear approximation of the triangle vertices. For this, we determine the triangle the closest point $\mathrm{clp}_{\mathcal{M}}(p)$ lies in when projected along its corresponding normal vector. Other approaches of higher order are possible.

B-spline basis

The B-spline basis for the approximation only contains those elements which have a non-empty intersection with the domain Ω:

$$\mathcal{B} := \left\{ k \in \mathbb{Z}^d \mid \sigma_k \cap \Omega \neq \emptyset \right\}.$$

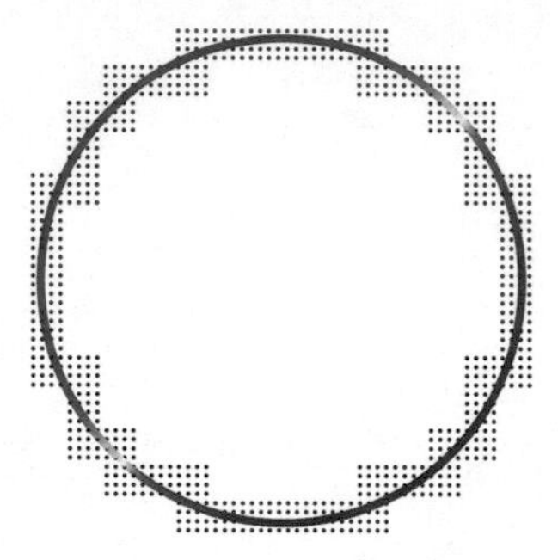

Figure 4.6: All data sites $\mathcal{P}$ within the corresponding active cells.

Analogously to the cells, we call a B-spline *active* if its corresponding index is contained in the index set $\mathcal{B}$. The function space of the treated splines is denoted by $\mathcal{S}_\Omega$ indicating that an evaluation is only reasonable on the domain Ω.

Discrete least-squares approximation
We use a discrete least-squares approximation as already indicated through the introduction of the data sites. We seek for a spline function S which minimizes the least squares residuals at the data sites:

$$\min_{S\in\mathcal{S}_\Omega} \sum_{i=1}^{|\mathcal{P}|} \|S(p_i) - G(p_i)\|^2, \quad p_i \in \mathcal{P}.$$

We assure that there are always more than or as many data sites as degrees of freedom because each cell is filled with n^d data sites and covered by n^d active B-splines. Although one has local optimal approximation power for tensor product B-splines for continuous approximations as introduced in Section 3.2, in order to transfer these results reasonably into the discrete setting, one needs to assure some properties of the underlying data to avoid large boundary artifacts. As elaborated in [Pra09, DPR13], besides the stability of the used basis, one has to restrict the so-called fill distance which corresponds to the biggest radius of a ball in a cell not containing one of the data sites. The even distribution of the data sites guarantees the fill distance to be $\frac{\sqrt{d}h}{2n}$ as well as the boundedness of the number of data sites in each grid cell. The global Matrix M for the least-squares fit consists of entries

$$M_{i,j} = B_{k(j)}(p_i),$$

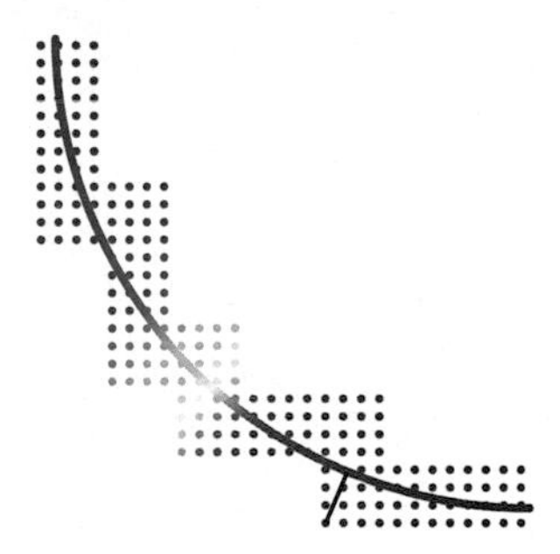

Figure 4.7: Illustration of the function value assignment: each data site is colored by the function of its closest point.

where $k(j)$ indicates that the index k refers to the jth basis element of the basis $\mathcal{B}$. The right-hand side of the linear systems is $G(p_i)$ for all rows i. We claim the minimizing problem to have a unique solution, i.e., in case the spline is supposed to be zero at all data sites, the corresponding coefficients should vanish. With the choice of our distribution of data sites, all the latter requirements are fulfilled. Therefore, the reasonable approximation results in the upcoming chapter are not surprising, cf. Section 4.3.

The $|\mathcal{P}| \times |\mathcal{B}|$ matrix of the scattered data problem can be stored in a sparse format. Each row of the matrix corresponds to one element of $\mathcal{P}$ but only has $n^d \ll |\mathcal{B}|$ non-zero entries. Additionally, due to the uniformity of the data sites distribution, it is sufficient to evaluate the B-splines on only one active cell and to simply copy these values for each cell to the corresponding active B-splines in $\mathcal{B}$.

Whereas the example for $d = 2$ leads to manageable matrix sizes and can therefore be solved globally, the matrices of the examples of the upcoming section for $d = 3$ may contain a large number of coefficients and are therefore numerically expensive. Consequently, the linear systems for $d = 3$ are solved by a local least-squares strategy, which is one possible quasi-interpolation. This means that we solve many small linear systems and only save those B-spline coefficients for which all data sites in the corresponding supports are completely considered in the local matrix.

In our implementation, all global and local linear systems are solved with the MATLAB© backslash operator. The computed spline coefficients b_k, for $k \in \mathcal{B}$, are in $\mathbb{R}^{\tilde{d}}$ if for the extended function holds $\mathcal{E}(g) : \Omega \mapsto \mathbb{R}^{\tilde{d}}$. Through this process, we obtain the spline $S \in \mathcal{S}_\Omega$

$$S : \mathbb{R}^d \supset \Omega \to \mathbb{R}^{\tilde{d}},$$
$$S(x) := \sum_{k \in \mathcal{B}} b_k B_k(x).$$

Finally, the desired spline s as approximant for the function g on the manifold $\mathcal{M}$ is obtained by the restriction of S to $\mathcal{M}$:

$$s : \mathcal{M} \to \mathbb{R}^{\tilde{d}},$$
$$s(x) := \mathcal{R}(S) = \sum_{k \in \mathcal{B}} b_k B_k(x). \quad x \in \mathcal{M}.$$

Figure 4.8 shows the error plot of the spline approximant with grid size $h = 0.15$, the biggest deviations from the function to be approximated occur unsurprisingly in the regions with steep derivatives.

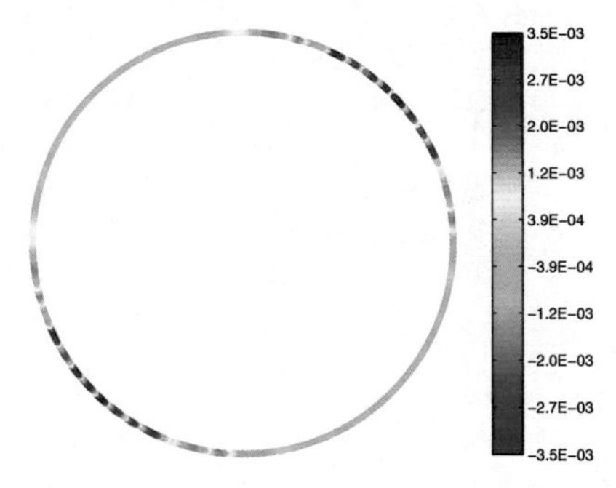

Figure 4.8: The unit circle is colored with the error of the spline approximation for the function $g(x,y) = \arctan(5(x-y))$ for $n = 4$ and $h = 0.15$.

4.3 Examples

The following section describes the approximation results of the ambient B-spline method of three different functions defined on the unit sphere. Whereas the first one treats a well-known example of a real function, the second and third actually represent surfaces of $\mathbb{R}^3$ whose single coordinate functions are interpreted as the functions to be approximated.

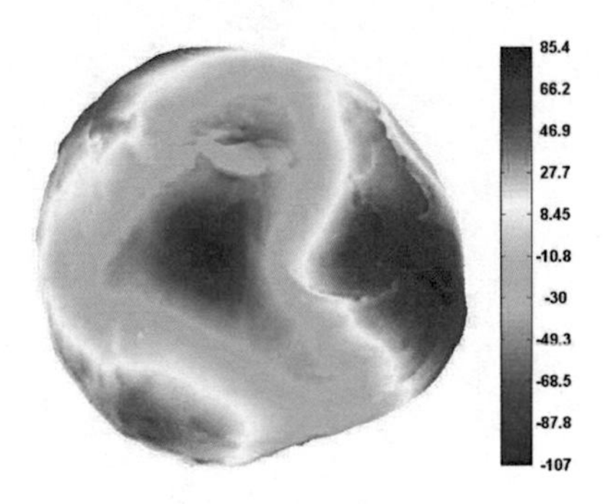

Figure 4.9: Exaggerated geoid undulations.

The first example we want to examine is significant within the field of geodesy and geophysics: the geoid. The geoid is the equipotential surface of the earth gravity field that best approximates mean sea level when tides, currents, winds etc. are neglected, see [KSV12]. In contrast to the geometrically idealized shape of the earth, the oblated spheroid or reference ellipsoid, the geoid is perpendicular to local gravity vectors, i.e. to the plumb lines, and therefore a reasonable choice as a datum for topographical heights, cf. [Van09]. For instance, a physically consistent reference surface has turned out to be meaningful for engineering applications, see [KSV12], such as the design of an aqueduct for which it is essential to know whether it diverts water up or down. The measurement of the strength of the gravitational field, the so-called gravimetry, became very accurate within the last decades due to the increasing success of satellite geodesy. The geoid height is the deviation of the geoid to the reference ellipsoid of the earth and varies by ± 100 m. The geoid's shape is mainly caused by anomalous density distributions within the earth. Figure 4.9 shows strongly exaggerated geoid undulations compared to the mean earth radius of about 6371 km because otherwise the bumps would be barely visible.

Figure 4.10 shows a cylindrical projection of the earth, colored by geoid heights and including its topographic relief. In some parts we notice a correlation between the topography and the geoid undulation, for example the Himalaya region with its massifs is closer to the center of the earth due to a stronger gravity whereas the Mariana Trench is farther away. Nevertheless, the geoid also has undulations that do not coincide with the obvious landmass, such as the region between Europe and Greenland.

Figure 4.10: Cylindrical earth projection with topographic heights and colored geoid undulation.

The following brief summary bases on explanations in [Sch05, Bar09, Gre03]. The geoid is often modeled by spherical harmonics, which are for all intents and purposes the analogon of Fourier series defined on the sphere and are also used outside geophysics or geodesy, for example within computer graphics, [Sch05] and [Gre03]. Spherical harmonics are eigensolutions of the angular part of the Laplacian and form a Schauder basis for the expansion of square-integrable functions on the sphere, a proof can be found in [AH12]. They are orthogonal functions and consist of associated Legendre polynomials $P_l^m(\cos(\theta))$, trigonometric functions $e^{im\phi}$ and a factor N_{lm}. The latter usually transforms them into an orthonormal set of functions. The angle ϕ refers to the longitude, the angle θ to the colatitude and the indices l and m are the band of the expansion and the approximation order, respectively. One defines

$$Y_l^m(\phi,\theta) := N_{lm} e^{im\phi} P_l^m(\cos(\theta))$$

to be the spherical harmonic function of degree l and order m. With this, one can model the Earth's gravitational potential W with a spherical harmonic truncated expansion

$$W(r,\phi,\theta) \approx \sum_{l=0}^{l_{\max}} \sum_{m=-l}^{l} W_{lm}(r) Y_l^m(\phi,\theta),$$

where W_{lm} are the coefficients to be determined, which also depend on the underlying radius r. One usually transforms the spherical harmonics into an orthogonal basis of real functions, but these details are not object of our considerations. There exist recurrence relations, which to some extend guarantee a numerically stable evaluation of spherical

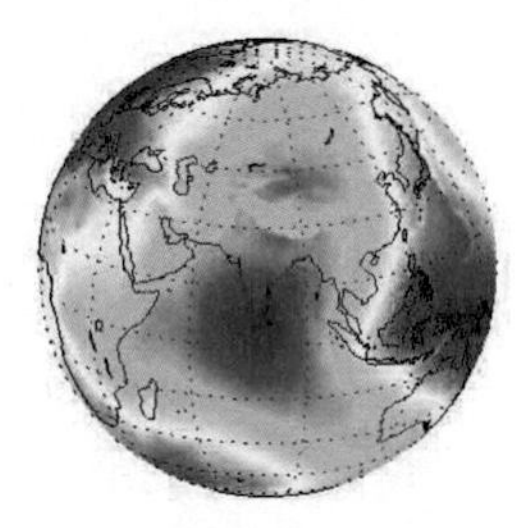

Figure 4.11: Geoid heights as colormap on the Earth taken from the MATLAB© function `geoidheight.m`.

harmonics. With this model, one of course has to use transformation formulas between the ellipsoidal and the spherical coordinates [Bar09].

Altogether, spherical harmonics are a good option for the approximation of functions on spheres but their technical details are substantial. Nevertheless, they hold two drawbacks: they are not a flexible tool for the approximation of functions on non-spherical manifolds and they have global support which means one has to evaluate the full set of the expansion in order to obtain a single function value. Contrarily, our approach of the ambient B-spline method overcomes both disadvantages and the technical details of the used function space are less complicated.

Thus, we are going to examine numerically how well our method can approximate geoid heights when we interpret these heights as function on the sphere. For this purpose we use the Earth Gravitational Model 1996, abbr. EGM96[1], as data basis, which consists of spherical harmonic coefficients up to degree and order 360 and arose of a collaboration of the NASA Goddard Space Flight Center and the National Imagery and Mapping Agency. The model consists of about $130,000$ coefficients and has an accuracy of 0.01m on a 15-minute grid of point values and a maximal worldwide error of ± 1m. Fortunately, this model is available in the Aerospace Toolbox of MATLAB© and we can therefore investigate how well we can approximate the geoid heights obtained by the MATLAB© function `geoidheight.m`. Since our approximation utilizes a grid within the ambient space whose points are in general not on the surface itself, the ambient B-spline method does not deliver

[1] `http://earth-info.nga.mil/GandG/wgs84/gravitymod/egm96/egm96.html`, accessed June 10, 2013

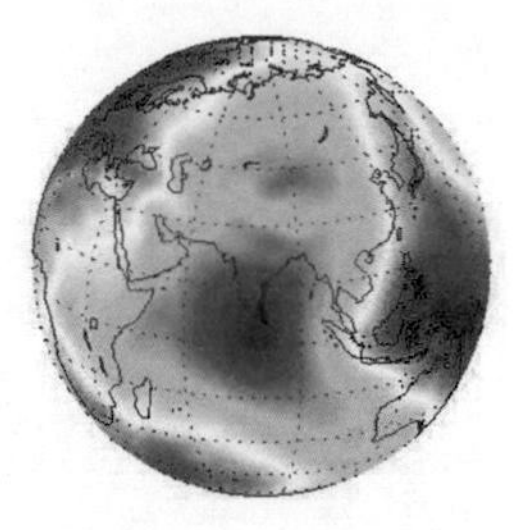

Figure 4.12: Geoid heights as colormap on the Earth taken from the ambient B-spline approximation with mesh size $h = 0.1$ and with about 7500 degrees of freedom.

coefficients which lead to a considerably better approximation on a certain subset of the sphere. We therefore strive for the goal of an overall approximation error of ± 1m, which we verify on a 15-minute grid. Although our approach, as expected, needs more coefficients and more input data for the approximation, once we have found an approximant, an evaluation of an arbitrary point on the the manifold only needs n^3 B-spline evaluations, where n is the B-spline order. Additionally, in case updated, measured data for the model in the region of, say, South America was available, a model based on spherical harmonics would have to be completely computed again whereas an approach with ambient B-splines could adapt the coefficients locally, i.e. within the region of interest.

We start by looking at an approximation with a very coarse mesh which leads to a clearly smaller number of degrees of freedom compared to the coefficients used in the EGM96. Figure 4.11 shows the geoid heights taken from the MATLAB© function and Figure 4.12 shows the approximated heights resulting from the ambient B-spline method.

We notice a smoother color gradient in Figure 4.12, which is due to the coarser resolution we have chosen for the approximation. A closer look at the error in Figure 4.13 displays the highly irregular distribution of the error which is already small in wide regions (green) but tremendously huge in others.

One possible course of action would be to decrease the mesh size until the approximation complies with the specified error threshold of ± 1m for the examined 15-minute grid. For instance, the bisection of the mesh size $h = 0.1$ results roughly in a bisection of the mean absolute error, but still has maximal absolute error of about 17m. Due to the

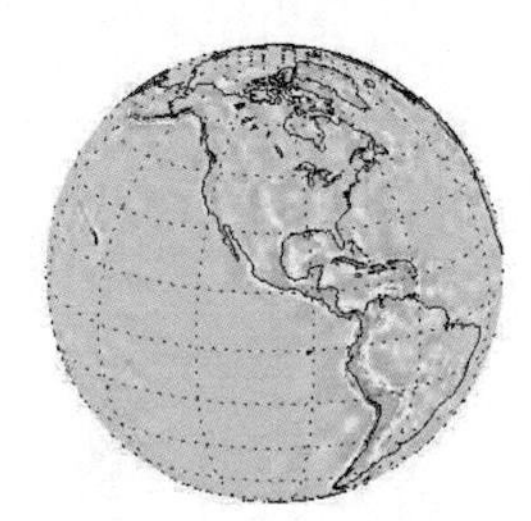

Figure 4.13: Errorplot reveals the geoid undulations to be highly irregular. Although huge parts of the surface correspond to small errors of less than 1m (green parts), the maximal absolute error on the 15-minute grid amounts to 22m for the B-spline approximation of mesh size $h = 0.1$

error estimate provided in Section 3.2, we would expect an error decrease of roughly $\left(\frac{1}{2}\right)^n$. We assume that our observation indicates that the chosen grid size is too coarse to assure the convergency order and we will come back to this point later. We also unnecessarily increase the degrees of freedom over broad regions, where the geoid undulations are already approximated well enough. Instead, we want to advance our approach with the help of adaptive refinement, which will be elaborated in the following chapter. In Chapter 6 we readdress the geoid as well as the upcoming examples of this section with the help of hierarchical overlays and B-splines and analyze the obtained approximation in detail.

4.3.2 Stanford Bunny

In the following, we address modeling examples of two-manifolds by interpreting each coordinate as function over a reference manifold with corresponding topology. The image of the three-dimensional spline approximant reveals the approximation of the target surface, if we evaluate the B-splines only on the reference manifold. Our next example is the well-known Stanford Bunny, which consists of scanned vertices taken from a clay figure and a generated triangulation. Due to the detailed structure of the suggested fur and holes at the bottom, it is a rather complicated model and therefore often chosen as test object for various algorithms within computer science. Nevertheless, we take the data provided by

Hoppe's webpage[2] and therefore use a patched model in which the holes on the bottom are artificially closed. With this, Praun and Hoppe may treat the model as a topological sphere and apply their spherical parametrization technique introduced in [PH03]. The bunny is shown in Figure 4.14 and the one-to-one mapping from the model to the unit sphere is illustrated in Figure 4.15.

Figure 4.14: Original mesh of the Stanford Bunny with 69630 triangles and 34817 vertices.

As already denoted in Section 2.1, the spherical parametrization is obtained by a coarse-to-fine optimization strategy which penalizes undersampling using a stretch-based measure, cf. [PH03]. This procedure copes with the challenges of highly deformed regions and prevents parametric foldovers, i.e. triangle flips, see Figure 4.16 in which we clearly recognize the distortion of the mapped ears. These properties are essential for the ambient B-spline method, especially the initialization of enough spherical area for high frequency surface detail guarantees that the grid size may remain reasonable.

We first test the bunny model with a relatively coarse grid size $h = 0.2$ and choose $n = 4$ to obtain a C^2-approximation. We have to admit that the mesh of the original clay figure only provides a continuous model and the generation of arbitrary smooth approximations may not be reasonable. However, since we later also investigate on the curvature properties of the results, we already choose a prescribed smoothness which lets us measure curvature

[2] `http://research.microsoft.com/en-us/um/people/hoppe/`, accessed May 2, 2013

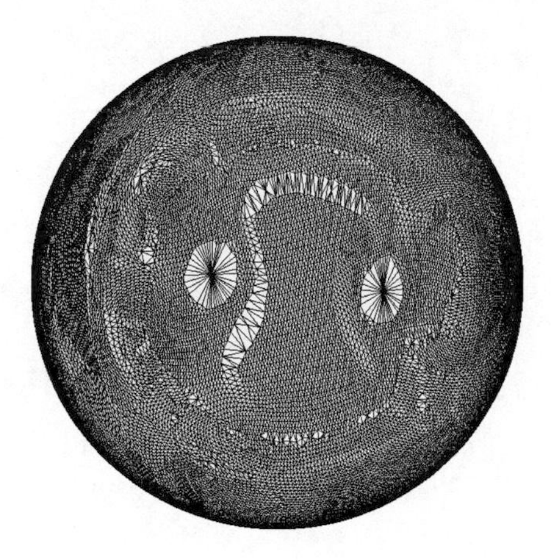

Figure 4.15: Vertices and corresponding triangles are mapped to sphere with the algorithm presented in [PH03]. This part of the sphere clearly reveals the stitched holes at the bottom of the bunny.

properties. It is not surprising that we obtain a spline approximant which does not reproduce details satisfactorily. Figure 4.17 illustrates the image of the original vertices mapped from the sphere colored by the least-squares error compared to the vertex positions of the original mesh. Since the given vertices are in general not part of the approximation, the least-squares error at these points provides a reasonable measure. The color plot of Figure 4.17 indicates that the chosen grid size has been too coarse.

Before decreasing the grid size, we point out that the high frequency parts of the ears are not only worst approximated but also reveal shape artifacts not contained in the model, cf. the close-up of the bunny's neck in Figure 4.18. These occur because the parametrization is highly distorted within this region and far too few B-splines treat the surface's details.

For the next approximation we bisect the grid size in each direction but keep the order of the B-spline approximation. The corresponding results are shown in Figure 4.19. The result demonstrates the significant error decrease, although especially the tips of the ears are still unsatisfactory. The number of active cells unsurprisingly roughly quadruples. We refer to ε_i as the difference of the ith vertex of the model to its approximation. The maximal least-squares error $\varepsilon_{\max} := \max_i \|\varepsilon_i\|_2 \approx 0.07$ and the root mean squares error

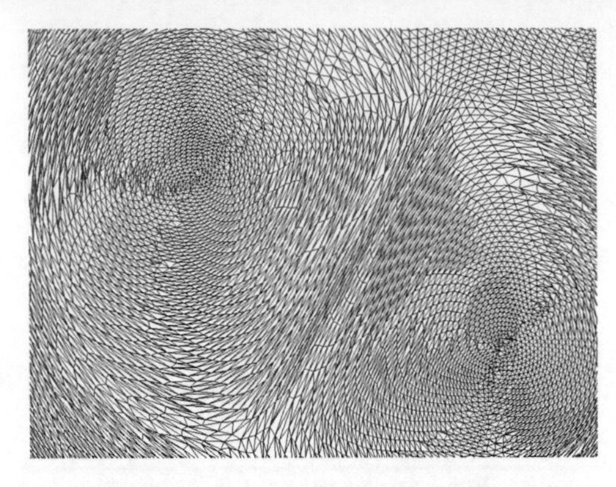

Figure 4.16: A close-up of the spherical parametrization of the Stanford Bunny showing the embedding of the bunny's ears.

$\varepsilon_{\text{rms}} := \sqrt{\frac{1}{\#\text{vertices}} \sum_i \|\varepsilon_i\|_2^2} \approx 0.0087$ for $h = 0.2$ drop to $\varepsilon_{\max} \approx 0.03$ and $\varepsilon_{\text{rms}} \approx 0.003$. Again, the error decrease does not comply with the estimate of Section 3.2 but here this is due to the generation of the function values which uses a linear approximation of the vertices of the corresponding triangles. Additionally, the close-up in Figure 4.20 reveals that shape artifacts can be avoided by choosing an appropriate grid size.

We notice that especially parts of the bunny's back are relatively well approximated even for the coarse grid size. This indicates that a global fine grid is not reasonable, particularly concerning computational costs. The head of the model tremendously differs from the shape of a sphere compared to the rest of the body. We again foreshadow the usage of a hierarchical spline space with which we can locally refine if necessary.

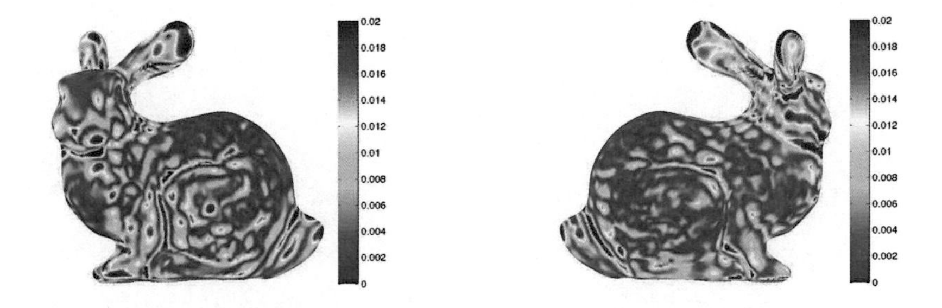

Figure 4.17: C^2-image of the original triangulation mapped from the sphere by the spline approximant of order 4 obtained with grid size $h = 0.2$. The B-spline basis contains 1784 B-splines and 416 active cells. Each vertex of the bunny is colored by its least-squares error $\|\varepsilon_i\|_2$.

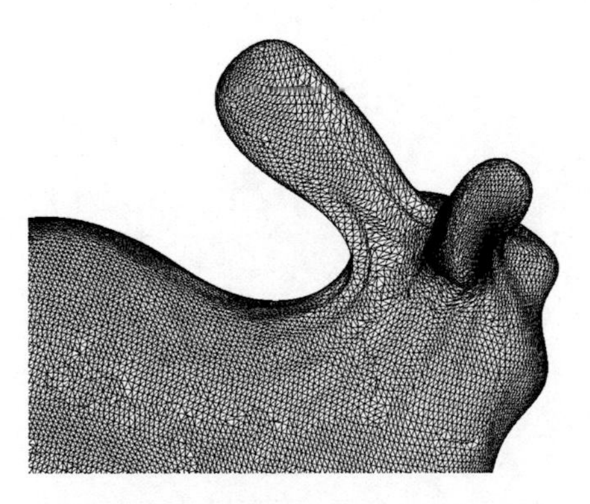

Figure 4.18: A close-up of the region of the bunny's neck reveals undesired waves caused by too coarse grid size.

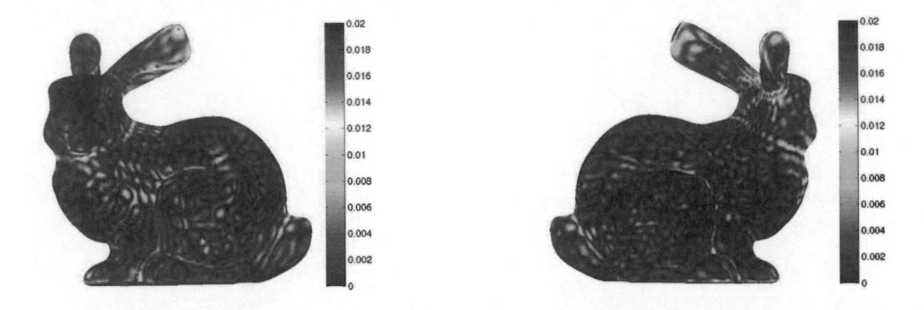

Figure 4.19: C^2-image of the original triangulation mapped from the sphere by the spline approximant of order 4 obtained with grid size $h = 0.1$. The B-spline basis contains 7411 B-splines and 1831 active cells. Each vertex of the bunny is colored by its least-squares error $\|\varepsilon_i\|_2$.

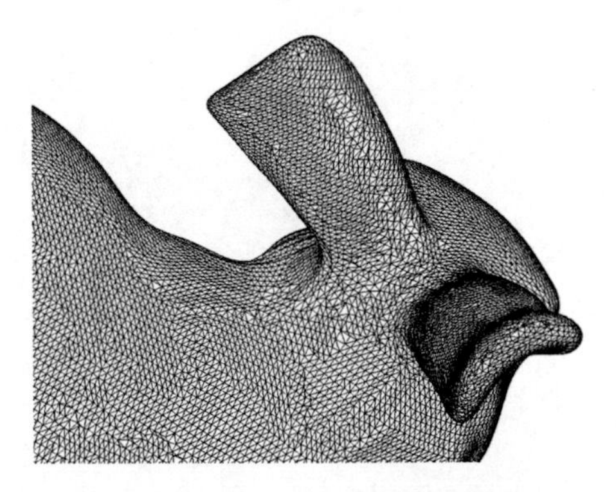

Figure 4.20: The close-up of the image generated with grid size $h = 0.1$ indicates that shape artifacts can be avoided by an appropriate resolution.

4.3.3 Cow model

We introduce a second surface example which is also taken from the data provided by Hoppe and Praun: a cow, see Figure 4.21. Although this model consists of roughly only one third of the vertices and triangles compared to the Stanford Bunny, it is to some extend more complicated since it contains more regions that significantly differ from a sphere. Note that the mesh of the cow's tail is only connected once to the cow's body so that we indeed deal with a topological sphere.

We naively start with an approximation with grid size $h = 0.2$ and order $n = 4$, and do not wonder about the undesired result illustrated in Figure 4.22. In contrast to the Stanford bunny, the cow is almost mirror symmetric when cut into two pieces along its back. That is why for now it suffices to look at the errors on one of the halves. The approximation fails to model legs, udder and details within the head and the tail, whereas the back is recovered comparatively well. This is allegeable because the coarse grid size prevents the function values of data sites to sample high frequency details and they are therefore not modeled. The image of the approximation obtained with the grid size $h = 0.08$ recovers more details, although it does not globally meet the requirement to reasonably model the cow's geometric attributes. For these two example computations, the maximal least-squares error at the cow vertices $\varepsilon_{\text{max}} \approx 0.074$ and the root mean squares error $\varepsilon_{\text{rms}} \approx 0.012$ for the grid size $h = 0.2$ drop to $\varepsilon_{\text{max}} \approx 0.037$ and $\varepsilon_{\text{rms}} \approx 0.0033$ for $h = 0.08$.

Finally, the cow model is indeed an excellent example for which hierarchical refinement leads to a desired approximation result without using unnecessarily many degrees of freedom because it contains distinct regions where a refinement is meaningful in order to dissolve shape details.

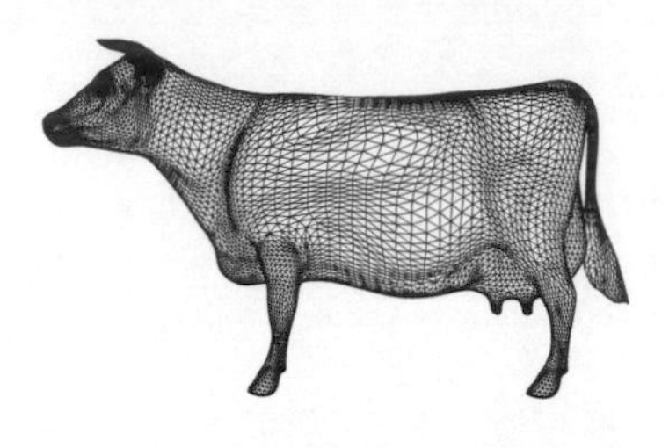

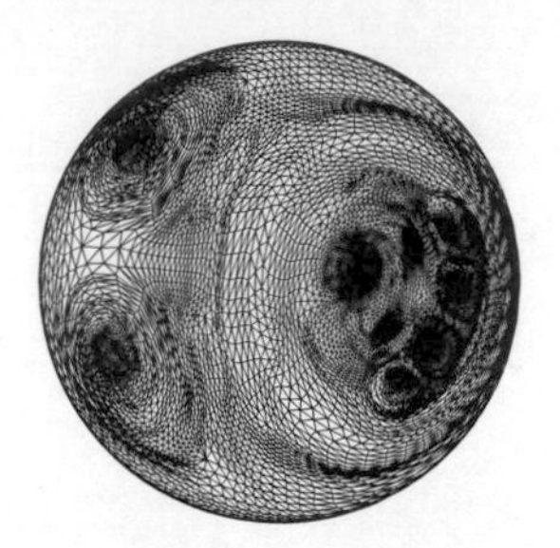

Figure 4.21: (Left): Original mesh of the cow's model with 23216 triangles and 11610 vertices.
(Right): The vertices and triangulations of the cow mapped to the unit sphere. We clearly recognize the regions of the two forelegs and the head.

Remark: As pointed out in [dB78], standard spline curves and surfaces in computer-aided design owe their popularity to the close relationship between the spline value and its B-spline coefficients. The *control points* provide a nice modeling tool since the movement of one control point causes a geometrically intuitive and local change of the curve or surface, respectively. One can even derive bounds for the distances between the spline curve or surface and its control points and show that the control points converge to the curve or surface, when successively decreasing the underlying grid size. Since we parametrize the example surfaces as splines $s : \mathbb{R}^3 \supset \mathcal{M} \to \mathbb{R}^3$, the control points lose the usual modeling property and cannot be easily used to modify the target object. Regarding the control points generated for the approximation of Figure 4.23, we apparently obtain an unstructured point cloud in $\mathbb{R}^3$, see Figure 4.24a. Nevertheless, neglecting some of the outer control points and zooming in towards the actual surface, we clearly observe a control point agglomeration along the target surface and recognize the shape of the cow, cf. Figure 4.24b. We assume that this control point concentration is caused by the almost degenerate third dimension of the spline approximation which forces the majority of the control points to align with the target surface.

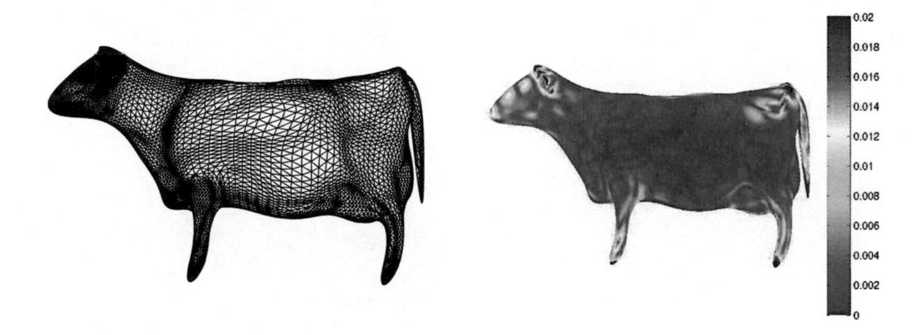

Figure 4.22: (Left): C^2-image of the original triangulation mapped from the sphere by the spline approximant of order 4 obtained with grid size $h = 0.2$. The B-spline basis consists of 1784 coefficients and 416 active cells. (Right): Colored (least squares) error plot.

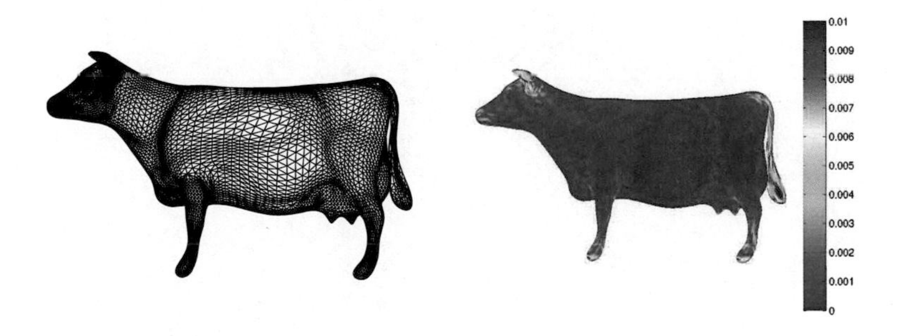

Figure 4.23: (Left): C^2-image of the original triangulation mapped from the sphere by the spline approximant or order 4 obtained with grid size $h = 0.08$. The B-spline basis consists of 11769 coefficients and 2846 active cells. (Right): Colored (least squares) error plot.

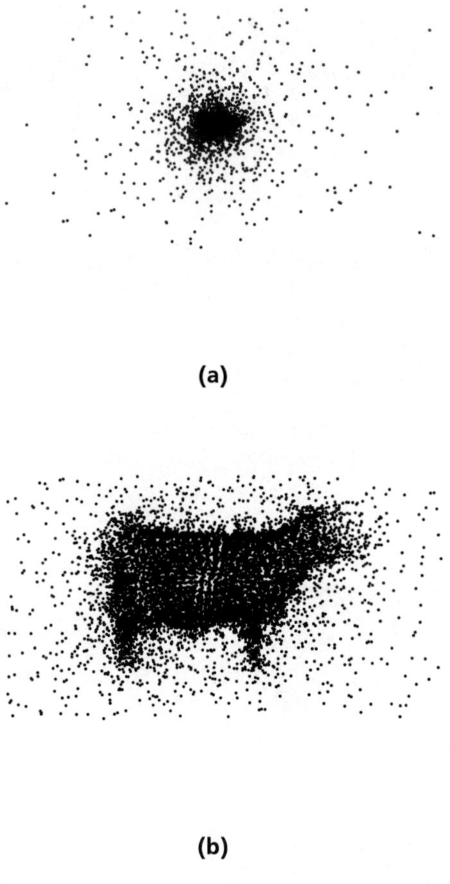

Figure 4.24: Corresponding control points in $\mathbb{R}^3$ of approximation example of Figure 4.23.

5 Hierarchical Approximation

The main benefit of hierarchical approaches as approximation technique is the usage of only as many degrees of freedom as needed by globally sticking to some error tolerance, see for example [FB88, FB95, Hor97, Kra98, SZBN03, VGJS11, DLP13]. We use hierarchical overlays as well as hierarchical B-splines in this thesis. The overlay approach is a well-known simple heuristic approach, and hierarchical B-splines have already been studied in detail by Kraft in [Kra98]. Nevertheless, he focuses on the approximation of tensor product surface patches and therefore only considers hierarchical B-splines in $\mathbb{R}^2$ with rectangular, nested domains as preimage. Since we want to use this idea within the ambient B-spline method in order to approximate functions on closed manifolds, we deal with different domains and specify the refinement process more restrictively.

5.1 Hierarchical grids and B-splines

We recapitulate the notations used in Section 4.2 and introduce the concept of orthogonal, dyadic grids we use for the determination of hierarchical B-splines. For this we establish the index ℓ which always refers to the refinement level.

Definition 5.1. *Let $\mathcal{G}_d^\ell$, $\ell \in \mathbb{N}_0$, be a disjoint partition of $\mathbb{R}^d$ into congruent d-dimensional boxes (cells), such that the origin of $\mathbb{R}^d$ is the lowest corner of some cell. Let $h \in \mathbb{R}_{>0}$ denote the isotropic grid size of $\mathcal{G}_d^0$, a cell $c_k^\ell \in \mathcal{G}_d^\ell, k \in \mathbb{Z}^d$ is defined as the half open box*

$$c_k^\ell := \left[\frac{k_1 h}{2^\ell}, \frac{(k_1+1)h}{2^\ell}\right) \times \cdots \times \left[\frac{k_d h}{2^\ell}, \frac{(k_d+1)h}{2^\ell}\right).$$

We associate to $\mathcal{G}_d^\ell$ the uniform, multivariate knot sequence

$$\mathcal{T}_d^\ell := h2^{-\ell}\mathbb{Z}^d.$$

The above definition means that any cell of the partition $\mathcal{G}_d^\ell$ comprises 2^d cells of $\mathcal{G}_d^{\ell+1}$. For applications we indeed need finite grids that cover our manifold and a sufficient neighborhood but for the interim introduction of our refinement rule, it is easier to deal with $\mathbb{R}^d$ as domain first.

We define the multivariate, uniform B-splines B_k^ℓ, again neglecting the isotropic order n, with knot sequence $\mathcal{T}_d^\ell$ and index $k = (k_1, \dots, k_d) \in \mathbb{Z}^d$ by

$$B_k^\ell(x) := \mathbf{B}_o^n\left(\frac{2^\ell}{h}x_1 - k_1\right) \cdot \ldots \cdot \mathbf{B}_o^n\left(\frac{2^\ell}{h}x_d - k_d\right).$$

Consequently, its support is the d-dimensional closed box

$$\operatorname{supp}\left(B_k^\ell\right) = \left[\frac{k_1 h}{2^\ell}, \frac{(k_1+n)h}{2^\ell}\right] \times \cdots \times \left[\frac{k_d h}{2^\ell}, \frac{(k_d+n)h}{2^\ell}\right].$$

Finally, we again define the half open box σ_k^ℓ of a B-spline B_k^ℓ by

$$\sigma_k^\ell := \left[\frac{k_1 h}{2^\ell}, \frac{(k_1+n)h}{2^\ell}\right) \times \cdots \times \left[\frac{k_d h}{2^\ell}, \frac{(k_d+n)h}{2^\ell}\right). \tag{5.1}$$

There are now two canonical scenarios for an approximation by hierarchical B-splines: the first is to use hierarchical overlays as suggested in [FB88, FB95, Hor97], the second is to generate a linearly independent hierarchical B-spline basis as for example suggested in [Kra98, VGJS11]. The first approach is easy to implement and needs no care for the extraction of a basis. However, it is numerically expensive and only provides a heuristic for the minimization of the least-squares error. We elaborate on the two variants because it turns out that both may have advantages for certain types of target functions.

5.2 Approximation by hierarchical overlays

We introduce a simple heuristic of an adaptive approach which has been considered before, cf. [FB88, FB95, Hor97]. The basic idea is first, to generate an initial approximation, second, to bisect the grid size, and third, to only treat those regions of the domain with an additional approximation which do not comply to the prescribed error tolerance. However, the second approximation does not use the original function, but the residuum remaining. The generation of overlays is continued until a given reference data $\mathcal{Y}$ on the manifold $\mathcal{M}$ complies to an error tolerance tol. Now and in the following sections, let g be a real function given on some manifold $\mathcal{M}$ and let $G = \mathcal{E}(g)$ be its extension to the manifold's ambient space as introduced in Chapter 3. We start by choosing an initial grid size h for the grid $\mathcal{G}_d^0$ and a B-spline order n. The first step of the hierarchical method works as follows:

- determine the domain of level 0

$$\Omega^0 := \left\{c_k^0 \in \mathcal{G}_d^0 \,|\, c_k^0 \cap \mathcal{M} \neq \emptyset\right\},$$

- determine all active B-splines of level 0

$$\mathcal{B}^0 := \left\{k \in \mathbb{Z}^d \,|\, \sigma_k^0 \cap \Omega^0 \neq \emptyset\right\},$$

- apply the ambient method of Section 4.2 with Ω^0, $\mathcal{B}^0$ and the function g to obtain the spline approximant $s^0 = \sum_{k\in\mathcal{B}^0} b_k^0 B_k^0$,
- determine the subset of all points violating the error tolerance:

$$\mathcal{K}^0 := \left\{y \in \mathcal{Y} \,|\, |s^0(y) - g(y)| > \mathrm{tol}\right\}.$$

If $\mathcal{K}^0$ is the empty set, the spline s^0 approximates the reference data with prescribed tolerance and no work remains. For the case that $\mathcal{K}^0 \neq \emptyset$, we may define an iterative process for upcoming levels. Assume $\mathcal{K}^{\ell-1}$ for $\ell > 0$ is not empty, then we implement the following procedure for level ℓ:

- determine the domain of level ℓ as the union of all cells which have a nonempty intersection with $\mathcal{K}^{\ell-1}$:

$$\Omega^\ell := \left\{c_k^\ell \in \mathcal{G}_d^\ell \,|\, c_k^\ell \cap \mathcal{K}^{\ell-1} \neq \emptyset\right\},$$

- determine all active B-splines of level ℓ

$$\mathcal{B}^\ell := \left\{k \in \mathbb{Z}^d \,|\, \sigma_k^\ell \cap \Omega^\ell \neq \emptyset\right\},$$

- apply the ambient B-spline method of Section 4.2 with Ω^ℓ and $\mathcal{B}^\ell$ and the function values

$$g^\ell(y) = g(y) - s^{\ell-1}(y), \quad y \in \mathcal{Y} \tag{5.2}$$

  in order to determine the spline coefficients of level ℓ and to finally obtain the spline approximant $s^\ell = \sum_{\tilde{\ell}=0}^{\ell} \sum_{k\in\mathcal{B}^{\tilde{\ell}}} b_k^{\tilde{\ell}} B_k^{\tilde{\ell}}$,
- determine the subset $\mathcal{K}^\ell$ of the reference data $\mathcal{Y}$ which still violates the error tolerance:

$$\mathcal{K}^\ell := \left\{y \in \mathcal{Y} \,|\, |s^\ell(y) - g(y)| > \mathrm{tol}\right\}.$$

- if $\mathcal{K}^\ell \neq \emptyset$ repeat the above steps for $\ell \leftarrow \ell + 1$.

Splines on higher levels cope with high frequencies of the function to be approximated. As already pointed out, this procedure may use unnecessary many degrees of freedom because each level is treated separately and does not produce a result in the least-squares sense when more than one level is used. The advantage is that one does not need to take care of linear dependencies or stability questions as we do in the upcoming section since all the least-squares problems are executed on only one level and afterward put together. We recalculate the example of the geoid with this approach in Section 6.1 because the highly irregular geoid function, cf. Section 4.3, does not have distinct connected parts where a refinement is especially reasonable in contrast to the example of the Stanford Bunny or the cow model.

5.3 Approximation by linearly independent hierarchical B-splines

We investigate on the second option more elaborately than on the heuristic of the previous section. The main idea of the usage of hierarchical B-splines is to replace B-splines with coarse knot sequences by B-splines with finer ones. The latter arise from the coarse B-splines through subdivision, see [Kra98]. Contrary to the heuristic of the previous section, the goal is to generate a linearly independent hierarchical B-spline basis in this process. Kraft established the theory for a basis selection on boxes in $\mathbb{R}^2$ which can be easily generalized for boxes in $\mathbb{R}^d$. A generalization to arbitrary domains is possible, see [Höl03], but in our setting we have to carefully define a hierarchical domain to avoid singularities in local least-squares matrices because every active cell of our domain is filled by n^d data sites but may be covered by more than n^d active, in $\mathbb{R}^d$ linearly independent, hierarchical B-splines. This may cause singular matrices when local strategies are used. Compared to the LR-splines introduced in [DLP13], hierarchical B-splines induce more degrees of freedom in favor of simpler grids and a less complex analysis of the mathematical framework.

First, we introduce the general refinement or subdivision rule in $\mathbb{R}^d$ without paying attention to restricted domains. It is well known that any univariate B-spline of order n defined on the knot sequence $\mathcal{T}_1^\ell$ has a unique expansion into B-splines defined on $\mathcal{T}_1^{\ell+1}$, cf. [Höl03] or [dB73]:

$$B_k^\ell = 2^{-(n-1)} \sum_{j=0}^{n} \binom{n}{j} B_{2k+j}^{\ell+1}.$$

For instance, if $n = 4$ we can express the univariate B-spline B_k^ℓ as follows:

$$B_k^\ell(x) = \frac{1}{8} B_{2k}^{\ell+1}(x) + \frac{1}{2} B_{2k+1}^{\ell+1}(x) + \frac{3}{4} B_{2k+2}^{\ell+1}(x) + \frac{1}{2} B_{2k+3}^{\ell+1}(x) + \frac{1}{8} B_{2k+4}^{\ell+1}(x),$$

see the illustration in Figure 5.1.

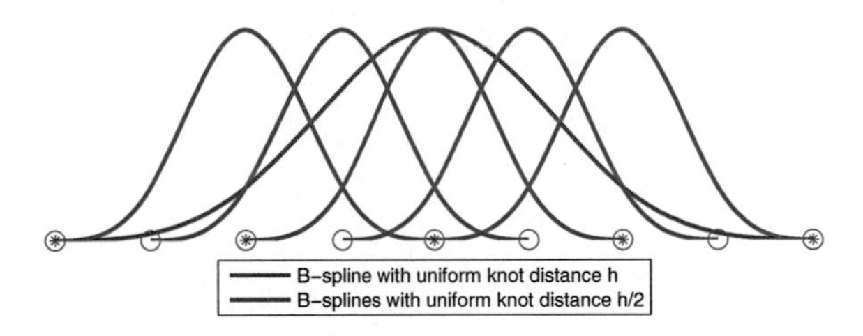

Figure 5.1: The univariate red-colored B-spline can be uniquely expressed in terms of the blue-colored B-splines.

For $d = 1$, the latter can be easily verified via knot insertion, see for example [dB78] or can be shown via induction, see [Höl03]. However, knot insertion for $d > 1$ results in an undesired global propagation of the refinement within the knot sequence, see [VGJS11]. Therefore, if we need to refine the B-spline B_k^ℓ of $\mathbb{R}^d$, we replace it by its $(n+1)^d$ unique counterparts on $\mathcal{T}_d^{\ell+1}$ like in [Kra98, VGJS11] and only add the knots that are necessary.

We again refer to a B-spline B_k^ℓ which belongs to the hierarchical basis as *active*. We therefore retain that the index k of the B-spline B_k^ℓ and its $(n+1)^d$ spanning counterparts on the next level must not be contained in the basis simultaneously or equivalently, not all of them can be active at once, in contrast to the heuristic of the previous section. Again, let $\mathcal{B}^\ell$ be the index set of all active B-splines with multivariate knot sequence $\mathcal{T}_d^\ell$. This means after refinement step L only $\mathcal{B}^0, \ldots, \mathcal{B}^L$ might be non-empty sets. In the following we always refer to L as the recent finest level.

We introduce two functions: the first function up assigns to an index set of level ℓ the indices of the above level to which the input is linear dependent. The second function down assigns to an index set of level ℓ all the indices of the level below which are already generated by the input indices:

Definition 5.2. *Let $\mathcal{K}^\ell$ be the index set of some active B-splines at level $\ell \geq 0$ with knot sequence $\mathcal{T}_d^\ell$. Then we refer to* $\mathrm{up}(\mathcal{K}^\ell)$ *as the index set of all B-splines with $\mathcal{T}_d^{\ell+1}$ that generate the linear span of $\{B_k^\ell | k \in \mathcal{K}^\ell\}$:*

$$\mathrm{up}(\mathcal{K}^\ell) := \{i \in \mathbb{Z}^d | i = 2k + m, k \in \mathcal{K}^\ell, m \in \{0,1,\ldots,n\}^d\}.$$

If $\ell > 0$, we define $\mathrm{down}(\mathcal{K}^\ell)$ *as the index set of all B-splines of level $\ell - 1$ that are in the linear span of $\{B_k^\ell | k \in \mathcal{K}^\ell\}$:*

$$\mathrm{down}(\mathcal{K}^\ell) := \{j \in \mathbb{Z}^d \,|\, \mathrm{up}(j) \subseteq \mathcal{K}^\ell\}.$$

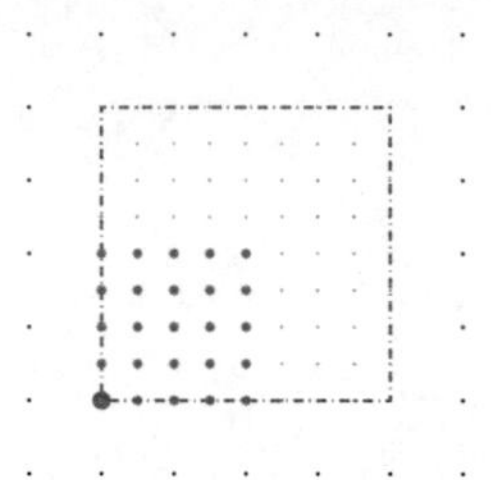

Figure 5.2: Illustration of the function up in $\mathbb{R}^2$ for order $n=4$: if k is the index of the B-spline in level ℓ we associate with the red knot • and which support is framed by -·-·-, then $\mathrm{up}(k)$ is the set of B-spline indices in level $\ell+1$ we identify with the blue knots $*$.
Conversely, in this trivial example it holds $\mathrm{down}(\mathrm{up}(k)) = k$.

We schematically illustrate the first refinement step and start by choosing an initial grid size h for $\mathcal{T}_d^0$, set $\mathcal{B}^0 = \mathbb{Z}^d$ and $\mathcal{B}^\ell = \emptyset$ for $\ell > 0$. Later we determine an index set of all B-splines that are not "good" (=fine) enough for sticking to some prescribed error tolerance. We refer to $\mathcal{K}^0$ as the index set of B-splines of level 0 which need to be refined. All these B-splines contain a point of our chosen reference data which is not approximated well enough compared to a given threshold. The index sets $\mathcal{K}^\ell$ for $\ell > 0$ are defined accordingly. The first refinement step is easiest described as follows:

- determine $\mathcal{K}^0$
- include all B-spline indices of level 1 that generate $\mathcal{K}^0$: $\mathcal{B}^1 = \mathrm{up}(\mathcal{K}^0)$
- remove all B-spline indices of level 0 that are linearly dependent of B-splines of level 1: $\mathcal{B}^0 \leftarrow \mathcal{B}^0 \setminus \mathrm{down}(\mathcal{B}^1)$.

Roughly speaking, we include active B-splines into the next hierarchical level with the index set generated by the function up and we exclude B-splines which are linearly dependent of the active B-splines just added a level above with the index set generated by the function down. Of course it always holds $\mathcal{K}^0 \subseteq \mathrm{down}(\mathcal{B}^1)$ but $\mathcal{K}^0$ can be a proper subset as it is illustrated in Figure 5.3. Technically, we would have to add the Index L to the sets $\mathcal{B}^{\ell,L}$ because the index set of active B-splines of level 0 for $L=0$ differs from the index set for $L=1$. However, we refer to the recent highest level L if necessary but neglect this index if it is clear that the sets may vary with increasing L.

So far we have the same strategy as presented in [Kra98], although we use a different notation. Nevertheless, we add a further rule for refinement steps $L>1$ which is due

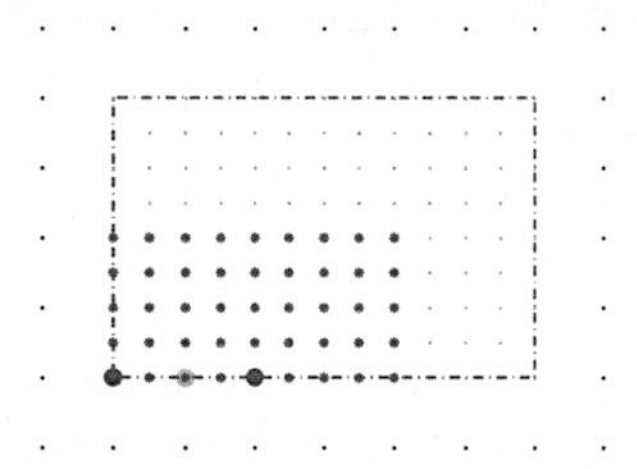

Figure 5.3: Illustration for $\mathcal{K}^0 \subsetneq \text{down}(\mathcal{B}^1)$ in $\mathbb{R}^2$ for order $n = 4$: if $\mathcal{K}^0$ refers to the two red knots • and which unified supports are framed by ----, then $\text{up}(\mathcal{K}^0)$ are the set of B-spline indices in level 1 we identify with the blue knots $*$. Contrary to the example in Figure 5.2, we obtain $\mathcal{K}^0 \subsetneq \text{down}(\text{up}(\mathcal{K}^0))$ because the B-spline associated to the cyan blue knot • is also linearly dependent of the B-splines which refer to the blue knots $*$.

to the domains we handle later: we forbid an overlapping of half open B-spline boxes, cf. Equation (5.1), which are more than one level apart. Its introduction not only simplifies the refinement process but also ensures the stability of the hierarchical basis, cf. Subsection 5.3.2. Thus, any further refinement step $L > 1$ has to assure that

$$\sigma_k^\ell \cap \sigma_m^{\ell+2} = \emptyset, \quad \forall k \in \mathcal{B}^\ell, \forall m \in \mathcal{B}^{\ell+2}, \forall \ell \in \{0,\dots,L-2\} \tag{5.3}$$

holds. Thus for general levels ℓ, we define the index set $\mathcal{J}^\ell \subseteq \mathcal{B}^\ell$ as the set of all B-splines at level ℓ which violate requirement (5.3):

$$\mathcal{J}^\ell := \{k \in \mathcal{B}^\ell \,|\, \exists m \in \mathcal{B}^{\ell+2} : \sigma_k^\ell \cap \sigma_m^{\ell+2} \neq \emptyset\},\ 0 \leq \ell \leq L-2.$$

Including requirement (5.3), we can generalize the refinement step of the first iteration to arbitrary steps $L > 0$ as follows:

Data: index set $\mathcal{K}^{L-1} \subseteq \mathcal{B}^{L-1}$ that needs to be refined
Result: hierarchical basis of $\mathbb{R}^d$ after refinement step L
include all B-spline indices of level L that generate $\mathcal{K}^{L-1}$: $\mathcal{B}^L = \text{up}(\mathcal{K}^{L-1})$;
initialize the index ℓ: $\ell = L-2$;
while $\ell \geq 0$ **do**
    determine the index set $\mathcal{J}^\ell$;
    determine all B-spline indices of level $\ell+1$ including the indices which may be added because of requirement (5.3) and remove the B-splines which are already generated by the level above:
    $\mathcal{B}^{\ell+1} \leftarrow (\mathcal{B}^{\ell+1} \cup \text{up}(\mathcal{J}^\ell)) \setminus \text{down}(\mathcal{B}^{\ell+2})$;
    if $\mathcal{J}^\ell$ *is empty* **then**
        $\ell = -1$;
    else
        $\ell \leftarrow \ell - 1$;
    end
end
remove all B-spline indices of level 0 that are linearly dependent to B-splines of the level above: $\mathcal{B}^0 \leftarrow \mathcal{B}^0 \setminus \text{down}(\mathcal{B}^1)$

Algorithm 1: General refinement rule in $\mathbb{R}^d$

The described course of action includes active B-splines in level L and removes linearly dependent B-splines of levels beneath with respect to the additional requirement (5.3). The refinement process and the impact of the additional requirement are illustrated in Figure 5.4 by means of a simple example in $\mathbb{R}^2$. The procedure generates linearly independent active B-splines in $\mathbb{R}^d$ by construction, cf. [Kra98]. We accept the increase of degrees of freedom through requirement (5.3) because it guarantees a stable basis for the approximation as is elaborated in Subsection 5.3.2. Of course the partition of unity gets lost in regions where two levels overlap. One could recover it via scaling, as suggested in [VGJS11]. Since such an approach depends on the underlying domain as already pointed out in [Kra98], we are not going to trace this idea because it would increase the computational complexity without a benefit for the approximation.

The generation of linearly independent B-splines in $\mathbb{R}^d$ or on boxes aligned with the knot sequences is fairly simple, cf. [Kra98] and the generalization for subsets of $\mathbb{R}^d$ is quickly done by only considering those B-splines that are relevant for the approximation, cf. [Höl03]. Since we deal with a discrete approximation, the difficulties in this thesis occur if we do not carefully define our hierarchical domain. The domain has to be within a tubular neighborhood of our reference manifold but we have to define the level of the active cells in the regions where two hierarchical B-spline levels overlap. Careless handling of the latter may lead to singular least-squares matrices although the B-splines are linearly inde-

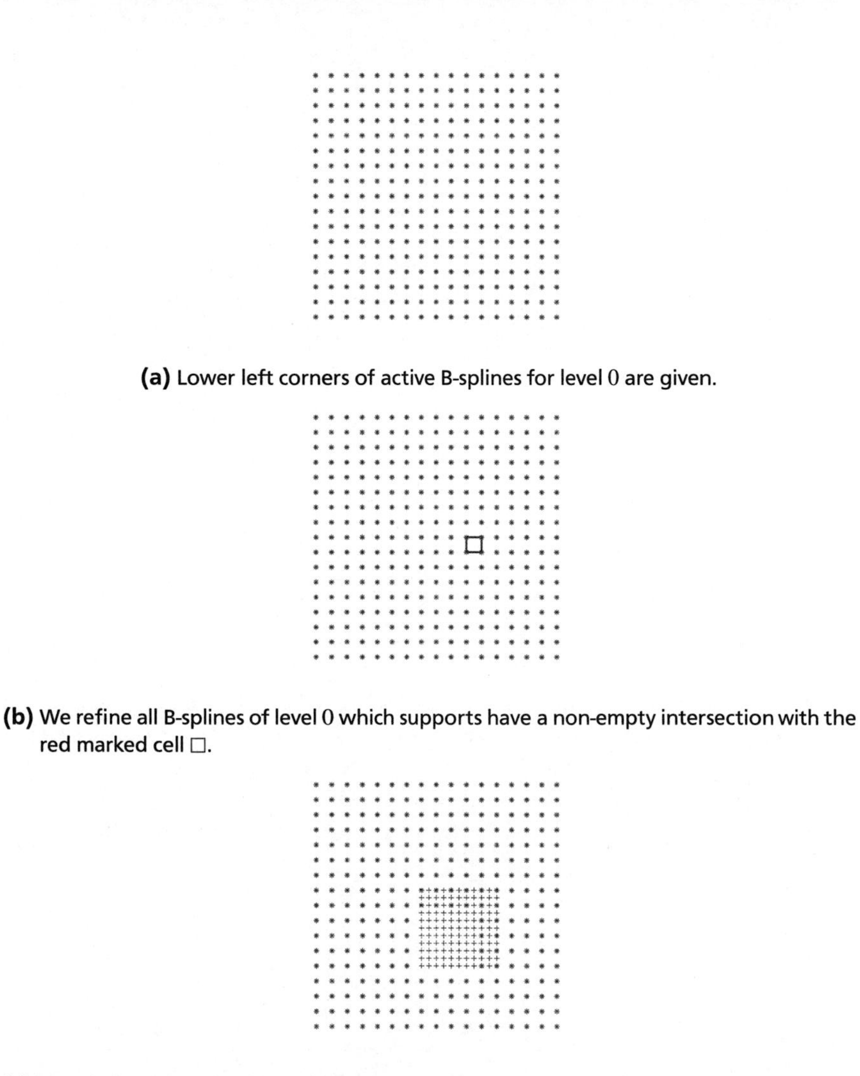

(a) Lower left corners of active B-splines for level 0 are given.

(b) We refine all B-splines of level 0 which supports have a non-empty intersection with the red marked cell □.

(c) Blue dots $*$ mark all active B-splines of level 0 and the red dots $+$ the active B-splines of level 1. All active B-splines are linearly independent.

Figure 5.4

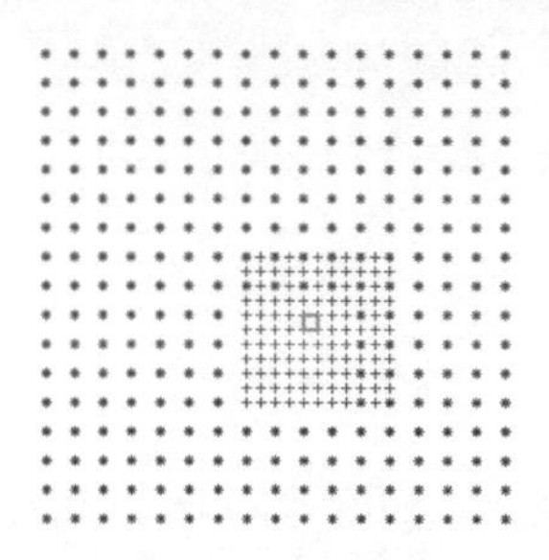

(d) We refine all B-splines of level 1 which supports have a non-empty intersection with the green marked cell □.

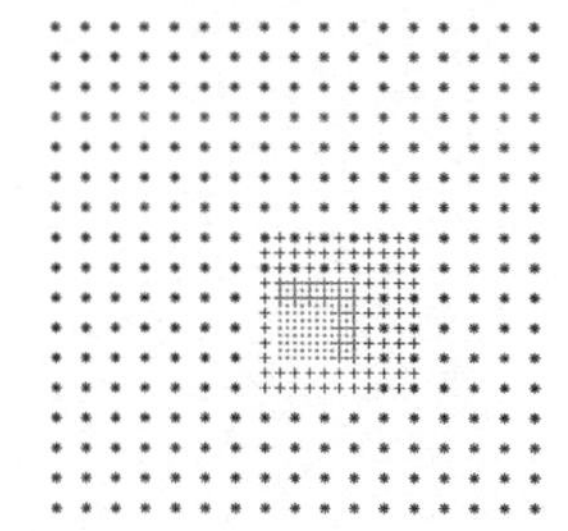

(e) Blue dots ∗ mark all active B-splines of level 0, the red dots + the active B-splines of level 1 and the green dots • the active B-splines of level 2. All active B-splines are linearly independent but requirement (5.3) is not fulfilled.

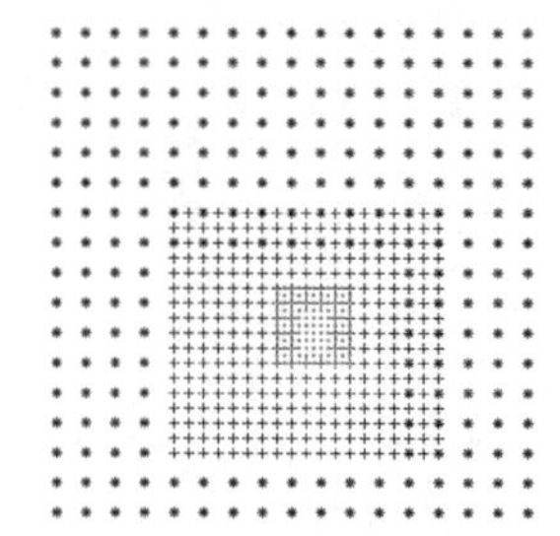

(f) Final result of the refinement. The active B-splines of the different levels are linearly independent.

Figure 5.4: Illustration of the refinement rule of Algorithm 1 in $\mathbb{R}^2$ for B-spline order $n = 4$.

pendent, cf. Figure 5.5. It turns out that a clever choice of the domain makes the extraction of a hierarchical B-spline basis together with the above considerations straightforward.

For this, we introduce how to establish a hierarchical domain which complies to an easy selection of linearly independent B-splines. We start by a definition of the domain on the current highest level L which consists of all cells of level L intersecting the manifold $\mathcal{M}$ and a half open B-spline box of the same level:

$$\omega^L := \{c_k^L \in \mathcal{G}_d^L \,|\, c_k^L \cap \mathcal{M} \neq \emptyset \wedge \exists m \in \mathcal{B}^L : \sigma_m^L \cap c_k^L \neq \emptyset\}.$$

The domains of the levels beneath are determined successively with the additional requirement that their cells must not have an intersection with any domain of higher levels:

$$\omega^\ell := \{c_k^\ell \in \mathcal{G}_d^\ell \,|\, c_k^\ell \cap \mathcal{M} \neq \emptyset \wedge \exists m \in \mathcal{B}^\ell : \sigma_m^\ell \cap c_k^\ell \neq \emptyset \wedge c_k^\ell \cap \omega^{\ell+1} = \emptyset\}, \quad \ell < L. \quad (5.4)$$

The complete domain for each approximation is simply obtained as union of the domains of the individual levels, $\Omega^L := \bigcup_{\ell \leq L} \omega^\ell$. Again, we actually would need to add the index L for the domains ω^ℓ since they are updated for each refinement step but disregard the index for the sake of convenience.

We impose one last additional requirement: the corresponding half open box of any active B-spline of level ℓ must intersect some active cell $c_k^\ell \in \omega^\ell$. In case there are any indices not fulfilling the latter, we additionally refine the corresponding indices. We therefore introduce the following set:

$$\mathcal{I}^\ell := \{k \in \mathcal{B}^\ell \,|\, \sigma_k^\ell \cap \omega^\ell = \emptyset\}.$$

Summing up, the refinement step L and the resulting recursive definition of the hierarchical domain and B-spline basis can be described as follows:

Data: manifold $\mathcal{M}$, hierarchical B-splines generated by Algorithm 1
Result: Ω^L and the hierarchical basis of Ω^L after refinement step L
for $\ell = L : -1 : 0$ **do**
  determine ω^ℓ;
  if $\ell < L$ **then**
    update B-spline indices according to (5.4) and with respect to the index set $\mathcal{I}^\ell$:
    $\mathcal{B}^{\ell+1} \leftarrow \mathcal{B}^{\ell+1} \cup \mathrm{up}(\mathcal{I}^\ell)$ and $\mathcal{B}^\ell \leftarrow \mathcal{B}^\ell \setminus \mathcal{I}^\ell$;
  end
end
$\Omega^L = \bigcup_{\ell=0}^L \omega^\ell$;
for $\ell = L : -1 : 0$ **do**
  final update of B-spline indices: $\mathcal{B}^\ell \leftarrow \{k \in \mathcal{B}^\ell \,|\, \sigma_k^\ell \cap \omega^\ell \neq \emptyset\}$;
end

Algorithm 2: Generation of hierarchical domain and corresponding B-spline basis.

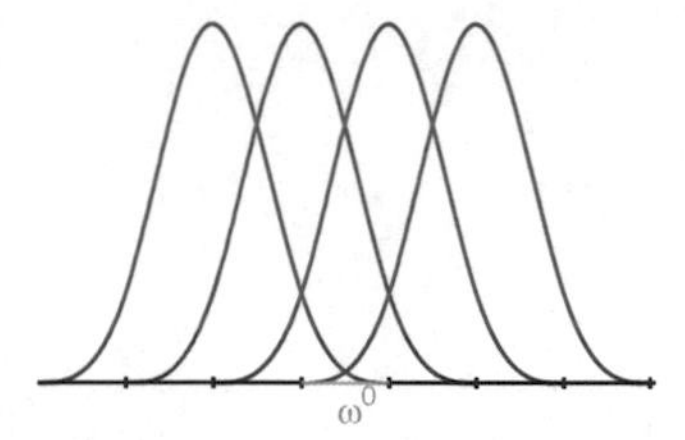

(a) All B-splines of level 0 that are relevant to approximate on $\omega^0 = \Omega^0$. Suppose the cubic, red-colored B-spline has to be refined.

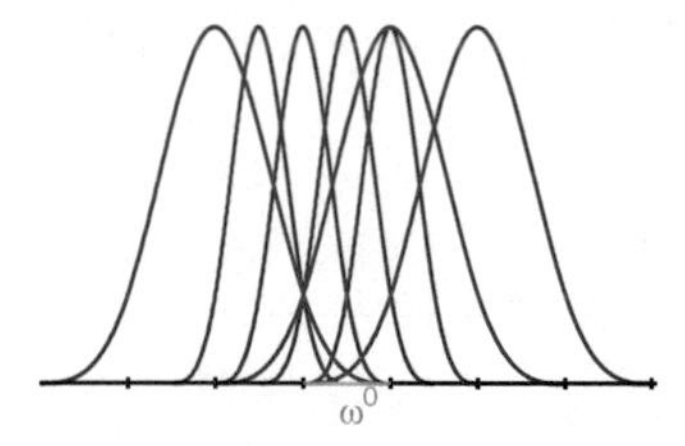

(b) These B-splines of different hierarchy levels are linearly independent in $\mathbb{R}$ but dependent on ω^0 because on each of the two cells of level 1 contained in ω^0 there are maximal $n = 4$ linearly independent polynomials.

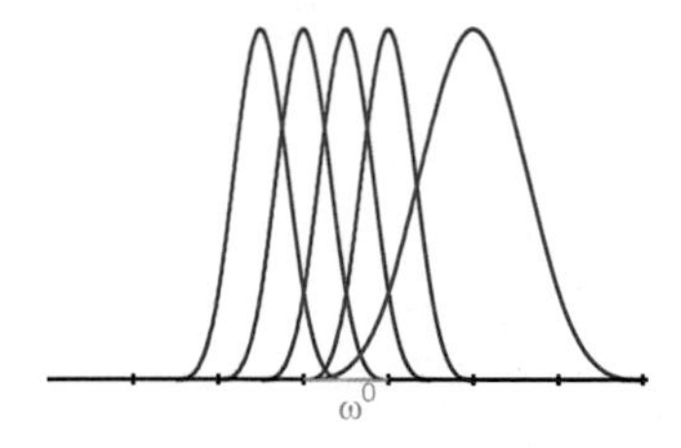

(c) These B-splines present a linearly independent basis for the interval ω^0 when keeping the finest B-splines. If $\Omega^1 = \omega^0$, i.e., the interval is still treated as a cell of level 0, it is filled by only four data sites although we have to determine five B-spline coefficients.

Figure 5.5: Simple example.

For an illustration of the hierarchical procedure, Figure 5.6 shows the hierarchical active cells of the recomputed example introduced in Section 4.2. The Algorithm detects regions where the function has large higher order derivatives and enforces finer grids for the B-splines to obtain an improved approximant.

Finally, we can define the hierarchical spline space $\mathcal{S}^{\mathrm{H}}_{\Omega^L}$ as the linear span of all active B-splines

$$\mathcal{S}^{\mathrm{H}}_{\Omega^L} := \operatorname{span}\{B^\ell_k \,|\, 0 \le \ell \le L \text{ and } k \in \mathcal{B}^\ell\}$$

and therefore obtain the representation of a spline function $S \in \mathcal{S}^{\mathrm{H}}_{\Omega^L}$ as follows:

$$\begin{aligned} S : \mathbb{R}^d \supset \Omega^L &\to \mathbb{R}^{\tilde{d}}, \\ x \mapsto S(x) &= \sum_{\ell \le L} \sum_{k \in \mathcal{B}^\ell} b^\ell_k B^\ell_k(x), \quad b^\ell_k \in \mathbb{R}^d. \end{aligned} \tag{5.5}$$

Each $x \in \Omega^L$ is only contained in less than $2n^d$ half open boxes of active B-splines due to requirement (5.3).

In standard spline theory, one may show the linear independence of the B-splines with the help of Marsden's identity [dB78]. We are able to prove the linear independence of the hierarchical selection of Algorithm 2 with the help of this result and induction on ℓ. The proof is straightforward due to the structure of the hierarchical domain and the B-spline selection.

Theorem 5.3. *The B-splines $\{B^\ell_k \,|\, 0 \le \ell \le L,\, k \in \mathcal{B}^\ell\}$ generated by Algorithm 1 and 2 are linearly independent on Ω^L.*

Proof. We need to show that all spline coefficients vanish, if the spline is equal to zero on Ω^L:

$$S(x) = \sum_{\ell=0}^{L} \sum_{k \in \mathcal{B}^\ell} b^\ell_k B^\ell_k(x) \equiv 0 \Rightarrow b^\ell_k = 0 \quad \text{for arbitrary } k, \ell.$$

Without loss of generality, we assume that $\mathcal{B}^0 \neq \emptyset$. Otherwise we start with the first level whose corresponding index set is not empty. We restrict S to ω^0, which is a nonempty set. We know that all B-splines of higher levels are zero because we require that an active cell of level ℓ cannot intersect the half open box of an active B-spline of level $\ell+1$, cf. (5.4). Neglecting the argument, this leads to

$$S|_{\omega^0} = \sum_{k \in \mathcal{B}^0} b^0_k B^0_k = 0.$$

From standard B-spline theory we know that the B-splines $\{B^0_k\}$ are linearly independent and conclude therefore that all coefficients $\{b^0_k \,|\, k \in \mathcal{B}^0\}$ are zero if $S \equiv 0$. Using this as

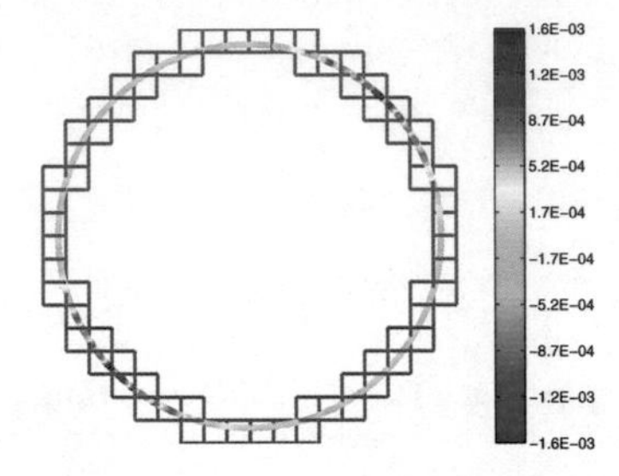

(a) The function $f(x,y) = \arctan(5(x-y))$ on the circle is approximated by a spline with grid size $h = 0.12$ and order $n = 4$. The active cells in blue □ belong to level 0, $\Omega^0 = \omega^0 = \bigcup \square$.

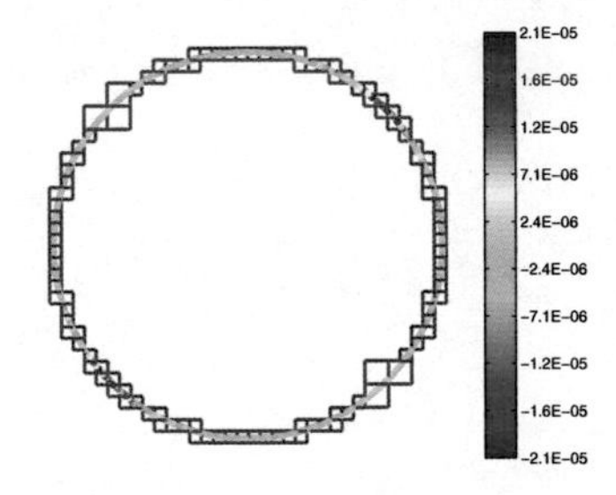

(b) After the first refinement we obtain the blue active cells of level 0 □ which define ω^0 and the red active cells of level 1 □ which define ω^1, $\Omega^1 = \omega^0 \cup \omega^1$.

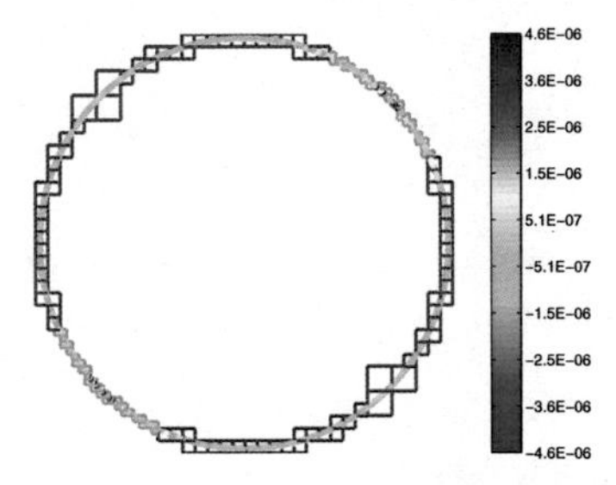

(c) After the second refinement we obtain the blue active cells of level 0 □, the red active cells of level 1 □ and the green active cells of level 2 □, $\Omega^2 = \omega^0 \cup \omega^1 \cup \omega^2$.

Figure 5.6: Recomputed example of Section 4.2 with error tolerance of $2 \cdot 10^{-5}$.

base case, the induction hypothesis is that all active B-splines up to level $\ell-1$ are linearly independent. If $\mathcal{B}^\ell \neq \emptyset$, we restrict the spline S to ω^ℓ, on which all B-splines of level greater than ℓ are zero. Using the induction hypothesis, we know that all B-splines up to level $\ell-1$ are zero, if $S \equiv 0$. This leads to

$$S|_{\omega^\ell} = \sum_{k\in\mathcal{B}^\ell} b_k^\ell B_k^\ell = 0.$$

Standard spline theory guarantees again the linear independence of the B-splines $\{B_k^\ell\}$ and we can therefore deduce that all $\{b_k^\ell \,|\, k \in \mathcal{B}^\ell\}$ are zero, too. This finishes the proof because $0 \leq \ell \leq L$ was chosen arbitrarily. □

5.3.1 Hierarchical quasi-interpolant

Our goal is to verify the stability of the hierarchical B-spline basis constructed by Algorithms 1 and 2. For the standard tensor product B-splines the tool used for the proof is the de Boor-Fix functional [dBF73] introduced in Section 4.1. We adapt this approach to our basis, introduce a hierarchical quasi-interpolant analogously to [Kra98] and verify some of the properties we actually need to show stability in Section 5.3.2. We use slightly different notations due to the different structure of our domain and neglect again the function or B-spline arguments to improve readability.

Definition 5.4. *Let λ^ℓ be the quasi-interpolant of order n as introduced in Section 4.1 for all B-splines with knot sequence $\mathcal{T}_d^\ell$ with the restriction that $k \in \mathcal{B}^\ell$ and λ_k^ℓ evaluates G (smooth enough or approximated by a smooth function) only on active cells of level ℓ which intersect σ_k^ℓ, i.e. on $\omega^\ell \cap \sigma_k^\ell$:*

$$\lambda^\ell : G|_{\omega^\ell} \mapsto \lambda^\ell G = \sum_{k\in\mathcal{B}^\ell} \left(\lambda_k^\ell G\right) B_k^\ell.$$

We define a hierarchical quasi-interpolant Q^L as follows:

$$\begin{aligned} Q^L : G \mapsto Q^L G &= Q^{L-1}G + \lambda^L(\mathrm{id} - Q^{L-1})G, \quad L > 0, \quad Q^0 = \lambda^0, \\ &= \sum_{\ell\leq L}\sum_{k\in\mathcal{B}^\ell} \left(Q_k^\ell G\right) B_k^\ell. \end{aligned} \tag{5.6}$$

We derive $Q_k^\ell = \lambda_k^\ell\left(\mathrm{id} - Q^{\ell-1}\right)$ from Equation (5.6) which basically means, that on each level ℓ one computes all B-spline coefficients of $\mathcal{B}^\ell$ with the standard quasi-interpolant for the difference of the function G and the already computed spline coefficients of all previous levels, see [Kra98]. Analogously to Kraft, we verify the following properties of the constructed hierarchical quasi-interpolant:

Lemma 5.5. *The hierarchical quasi-interpolant Q^L of Definition 5.4*
***a)** reproduces polynomials of coordinate order n:*

$$Q^L \pi = \pi \quad \textit{for all polynomials } \pi \in \Pi_n(\Omega^L),$$

***b)** fulfills*

$$Q_k^\ell S = b_k^\ell \quad \textit{for } 0 \leq \ell \leq L \quad \textit{and} \quad k \in \mathcal{B}^\ell,\ S \in \mathcal{S}_{\Omega^L}^{\mathrm{H}},$$
$$Q_k^\ell B_m^{\tilde{\ell}} = \delta_{\ell\tilde{\ell}} \delta_{km} \quad \textit{for } 0 \leq \tilde{\ell}, \ell \leq L \quad \textit{and} \quad m \in \mathcal{B}^{\tilde{\ell}},\ k \in \mathcal{B}^\ell,$$

***c)** satisfies the projection property:*

$$Q^L Q^L G = Q^L G.$$

Proof. **a)** Neglecting the arguments of π and the B-splines, we show via induction on ℓ that

$$Q^\ell \pi = \pi - \sum_{\ell+1 \leq \tilde{\ell} \leq L} \sum_{k \in \mathcal{B}^{\tilde{\ell}}} b_k^{\tilde{\ell}} B_k^{\tilde{\ell}} \quad \text{for some } b_k^{\tilde{\ell}} \in \mathbb{R}^d \text{ and } 0 \leq \ell \leq L$$

holds. Since we only deal with finitely many hierarchical levels, Lemma 5.5**a)** is a consequence of the latter.

Without loss of generality, throughout the proof of **a)** and **b)** we assume $\mathcal{B}^0 \neq \emptyset$. For the base case $\ell = 0$ we verify for any polynomial $\pi \in \Pi_n(\Omega^L)$:

$$Q^0 \pi = \lambda^0 \pi = \sum_{k \in \mathcal{B}^0} \left(\lambda_k^0 \pi\right) B_k^0 = \pi - \sum_{\{k \in \mathbb{Z}^d \setminus \mathcal{B}^0 \,|\, \sigma_k^0 \cap \Omega^L \neq \emptyset\}} b_k^0 B_k^0. \tag{5.7}$$

Regarding (5.7), we know that λ^0 reproduces polynomials. Therefore we would obtain π on Ω^L if all relevant B-splines of level 0 were active. If we have more than one level, we have to subtract the non-active B-splines that are relevant for Ω^L, which can be expanded by

$$\sum_{\{k \in \mathbb{Z}^d \setminus \mathcal{B}^0 \,|\, \sigma_k^0 \cap \Omega^L \neq \emptyset\}} b_k^0 B_k^0 = \sum_{1 \leq \ell \leq L} \sum_{k \in \mathcal{B}^\ell} b_k^\ell B_k^\ell$$

for some b_k^ℓ. The induction hypothesis reads as follows:

$$Q^{\ell-1} \pi = \pi - \sum_{\ell \leq \tilde{\ell} \leq L} \sum_{k \in \mathcal{B}^{\tilde{\ell}}} b_k^{\tilde{\ell}} B_k^{\tilde{\ell}}.$$

Finally, we get for Q^ℓ:

$$\begin{aligned}
Q^\ell \pi &= Q^{\ell-1}\pi + \lambda^\ell(\pi - Q^{\ell-1}\pi) \\
&= \pi - \sum_{\ell \le \tilde{\ell} \le L} \sum_{k \in \mathcal{B}^{\tilde{\ell}}} b_k^{\tilde{\ell}} B_k^{\tilde{\ell}} + \lambda^\ell \sum_{\ell \le \tilde{\ell} \le L} \sum_{k \in \mathcal{B}^{\tilde{\ell}}} b_k^{\tilde{\ell}} B_k^{\tilde{\ell}} \\
&= \pi - \sum_{\ell \le \tilde{\ell} \le L} \sum_{k \in \mathcal{B}^{\tilde{\ell}}} b_k^{\tilde{\ell}} B_k^{\tilde{\ell}} + \sum_{k \in \mathcal{B}^\ell} b_k^\ell B_k^\ell \\
&= \pi - \sum_{\ell+1 \le \tilde{\ell} \le L} \sum_{k \in \mathcal{B}^{\tilde{\ell}}} b_k^{\tilde{\ell}} B_k^{\tilde{\ell}},
\end{aligned}$$

where we used the induction hypothesis in the second row and the property that the standard quasi-interpolant λ^ℓ evaluates the spline only on ω^ℓ in the third row.
b) We show this property again via induction on ℓ. For $S \in \mathcal{S}^{\mathrm{H}}_{\Omega^L}$ and $\ell = 0$ we get the base case:

$$Q_k^0 S = \lambda_k^0 S = \lambda_k^0 \left(\sum_{m \in \mathcal{B}^0} b_m^0 B_m^0 + \sum_{1 \le \ell \le L} \sum_{m \in \mathcal{B}^\ell} b_m^\ell B_m^\ell \right) = b_k^0,$$

because the functional λ_k^0 evaluates the spline only on ω^0 on which the second summand vanishes. Using the induction hypothesis $Q_k^{\ell-1} S = b_k^{\ell-1}$, we obtain:

$$\begin{aligned}
Q_k^\ell S &= \lambda_k^\ell \left(s - Q^{\ell-1} s \right) \\
&= \lambda_k^\ell \left(\sum_{0 \le \tilde{\ell} \le L} \sum_{m \in \mathcal{B}^{\tilde{\ell}}} b_m^{\tilde{\ell}} B_m^{\tilde{\ell}} - \sum_{0 \le \tilde{\ell} \le \ell-1} \sum_{m \in \mathcal{B}^{\tilde{\ell}}} b_m^{\tilde{\ell}} B_m^{\tilde{\ell}} \right) \\
&= \lambda_k^\ell \left(\sum_{\ell \le \tilde{\ell} \le L} \sum_{m \in \mathcal{B}^{\tilde{\ell}}} b_m^{\tilde{\ell}} B_m^{\tilde{\ell}} \right) \\
&= b_k^\ell,
\end{aligned}$$

where we used the same argument for λ_k^ℓ as for λ_k^0 in the third row. For the special case $S := B_m^{\tilde{\ell}}$ we conclude

$$Q_k^\ell B_m^{\tilde{\ell}} = \delta_{\ell\tilde{\ell}} \delta_{km}.$$

c) Since $Q^L G \in \mathcal{S}^{\mathrm{H}}_{\Omega^L}$, it can be expressed as follows:

$$Q^L G = \sum_{0 \le \ell \le L} \sum_{k \in \mathcal{B}^\ell} b_k^\ell B_k^\ell.$$

Together with the previous considerations, we see that

$$\begin{aligned} Q^L Q^L G &= Q^L \left(\sum_{0\le\ell\le L} \sum_{k\in\mathcal{B}^\ell} b_k^\ell B_k^\ell \right) \\ &= \sum_{0\le\tilde{\ell}\le L} \sum_{m\in\mathcal{B}^{\tilde{\ell}}} Q_m^{\tilde{\ell}} \left(\sum_{0\le\ell\le L} \sum_{k\in\mathcal{B}^\ell} b_k^\ell B_k^\ell \right) B_m^{\tilde{\ell}} \\ \overset{\text{Lemma 5.4b)}}{=} & \sum_{0\le\tilde{\ell}\le L} \sum_{m\in\mathcal{B}^{\tilde{\ell}}} b_m^{\tilde{\ell}} B_m^{\tilde{\ell}} \\ &= Q^L G \end{aligned}$$

holds. □

Whereas Lemma 5.5**a)** and **c)** show that the constructed quasi-interpolant has desired properties, with the help of Lemma 5.5**b)** we may show the stability of the constructed hierarchical basis.

5.3.2 Stability

It is well-known that the standard tensor product B-spline basis is uniformly stable on $\mathbb{R}^d$. To guarantee this property for arbitrary domains is more complicated. In case the treated domain is a union of full grid cells, the basis fulfills the following estimate:

$$\text{const}(n,d) \max_k |b_k| \le \| \sum_k b_k B_k \|_\infty \le \max_k |b_k|$$

for some constant $\text{const}(n,d)$ which is independent of the knot sequence, see [dB78] or [Möß06]. Whereas the upper bound simply follows from partition of unity of the tensor product B-splines, one has to work harder to prove the lower bound. Fortunately, with our choice of hierarchical B-splines, the hierarchical domain of Section 5.3 and with the additional requirement (5.3), we are able to verify the stability of the basis constructed by Algorithms 1 and 2.

Theorem 5.6. *Let Ω^L be a hierarchical domain with corresponding hierarchical spline space $\mathcal{S}_{\Omega^L}^{\mathrm{H}}$ constructed as introduced in Section 5.3. There exist constants C_0 and C_1 which are independent of the underlying knot sequence and the number of levels such that for each $S \in \mathcal{S}_{\Omega^L}^{\mathrm{H}}$ holds:*

$$C_0(n,d) \max_{\substack{0\le\ell\le L \\ k\in\mathcal{B}^\ell}} |b_k^\ell| \le \| \sum_{0\le\ell\le L} \sum_{k\in\mathcal{B}^\ell} b_k^\ell B_k^\ell \|_{L^\infty(\Omega^L)} \le C_1 \max_{\substack{0\le\ell\le L \\ k\in\mathcal{B}^\ell}} |b_k^\ell|.$$

Sketch of Proof. The upper bound holds because of requirement (5.3): for each $x \in \Omega^L$ there are active B-splines of at most two different levels. Consequently, we may choose $C_1 = 2$ and obtain the second inequality.
For the lower bound, we need the help of the hierarchical quasi-interpolant of the previous section. For Q_k^ℓ we choose the evaluation point τ again to be in an active cell c_m^ℓ within the corresponding half open box, i.e. $c_m^\ell \cap \sigma_k^\ell$:

$$
\begin{aligned}
|Q_k^\ell S| &= |b_k^\ell| \\
&= |\lambda_k^\ell (S(\tau) - Q^{\ell-1} S(\tau))| = |\lambda_k^\ell \sum_{\tilde{k} \in \mathcal{B}^\ell} b_{\tilde{k}}^\ell B_{\tilde{k}}^\ell(\tau)| \\
&= \Bigg| \frac{1}{((n-1)!)^d} \sum_{0 \le j_1,\dots,j_d \le (n-1)} (-1)^{n-1-j_1} \cdots (-1)^{n-1-j_d} \cdots \\
&\quad \psi_{k_1}^{(n-1-j_1)}(\tau_1) \cdots \psi_{k_d}^{(n-1-j_d)}(\tau_d) \, \underbrace{\mathrm{D}^{(j)} \left(\sum_{\tilde{k} \in \mathcal{B}^\ell} b_{\tilde{k}}^\ell B_{\tilde{k}}^\ell(\tau) \right)}_{=: S^{(j)}(\tau)} \Bigg| \\
&\le \frac{1}{((n-1)!)^d} \sum_{0 \le j_1,\dots,j_d \le (n-1)} |\psi_{k_1}^{(n-1-j_1)}(\tau_1) \cdots \psi_{k_d}^{(n-1-j_d)}(\tau_d)| \, |S^{(j)}(\tau)| \\
&\le \frac{1}{((n-1)!)^d} \max_{0 \le j_1 \le (n-1)} |\psi_{k_1}^{(n-1-j_1)}(\tau_1)| \cdots \max_{0 \le j_d \le (n-1)} |\psi_{k_d}^{(n-1-j_d)}(\tau_d)| \cdots \\
&\quad \underbrace{\sum_{0 \le j_1,\dots,j_d \le (n-1)} |S^{(j)}(\tau)|}_{=: \|S\|_\tau} .
\end{aligned}
$$

Looking at the third row of the above inequalities, we recognize that the standard de Boor-Fix functional only acts on classic tensor product B-splines. As already mentioned in Section 4.1, this functional is invariant under affine transformations. With the help of a subtle transformation one obtains that the corresponding half open box σ_k^ℓ corresponds to the half open box $[0,1)^d$. Therefore τ lies within a cell $c \subset [0,1)^d$. The rest of the proof proceeds analogously to the classic result, e.g. [Möß06] for the univariate case: first we verify that each derivative of the univariate polynomials of Marsden's Identity is bounded by

$$
|\psi_{k_i}^{(n-1-j_i)}| \le (n-1)!.
$$

Second, we know that S is a polynomial on the cell c of Lebesgue measure $\frac{1}{n^d}$ and for polynomials, the defined sum $\|S\|_\tau$ is a norm on the finite-dimensional space $\Pi_n(c)$ and therefore equivalent to the norm $\|\cdot\|_{L^\infty(c)}$. The latter implies that there exists a constant $C(n,d)$ which satisfies the inequality

$$\|S\|_\tau \leq C(n,d)\|S\|_{L^\infty(c)}.$$

In conclusion, we obtain

$$|b_k^\ell| \leq C(n,d)\|S\|_{L^\infty(\Omega^L)},$$

and because the coefficient b_k^ℓ is chosen arbitrarily, we finally define $C_0(n,d) = C(n,d)^{-1}$. □

In summary, the stability of the hierarchical B-splines guarantees the numerical robustness of the linear systems to be solved using the ambient B-spline method.

5.3.3 Implementation aspects of hierarchical B-splines

One significant part of this thesis concerns the implementation of the introduced approach in MATLAB© and its verification by some examples. Without going into too many details, we want to comment on some of the difficulties and challenges concerning hierarchical refinement. The integration of the hierarchical B-splines of this section into the general ambient B-spline method introduced in Section 4.2 is an easy task and can be illustrated by Figure 5.7.

An efficient implementation for the hierarchical structure of the grids uses quadtrees for $\mathbb{R}^2$ or octrees for $\mathbb{R}^3$, respectively. The latter was pioneered by Meagher in [Mea80] and is simplest described as a tree data structure whose nodes are either leaves or contain usually exact eight children. For instance, it is extensively used for volume rendering, see the short survey [Kno06] for more details. Due to the decreasing domains, i.e. $\Omega^0 \supsetneqq \Omega^1 \supsetneqq \cdots \supsetneqq \Omega^L$, when hierarchical refinement is applied, we need a data structure which allows more than zero and less than 2^d children for a node.

Of course the function values of the data sites that belong to lower levels and remain unchanged under the refinement process can be reused. The determination of the least-squares matrices of the hierarchical approach is not as simple as introduced in Section 4.2 but can still be computed efficiently. One has to consider that maximal two hierarchy levels overlap, see (5.3), namely the data sites of an active cell of level ℓ are at most contained in B-spline supports of level ℓ and of level $\ell-1$. Therefore, it is sufficient to evaluate the n^d relevant B-splines on only one active cell of the same level as well as on a cell one level above. These two obtained base $n^d \times n^d$ matrices suffice to build the local or global matrices by assigning the corresponding active B-spline entries to the data sites.

By applying a local least-squares approach, the local matrices may be singular although the underlying B-splines form a basis for Ω^L. For instance, this may happen if B-splines of level ℓ but only active cells of levels $\ell+1$ and higher are part of the local linear system. This means that a linearly dependent column only may occur if not all active cells within its corresponding B-spline support are taken into account. There are two reasonable strategies to avoid this: one either only keeps or weights those B-spline coefficients of the local approximation for which all active cells within their supports are considered, or one starts by determining the B-spline coefficients of the lowest level and successively uses the coefficients of the previous level to obtain the upcoming. For the first suggestion, possible singularities of local systems do not cause any harm and for the second, singularities do not appear.

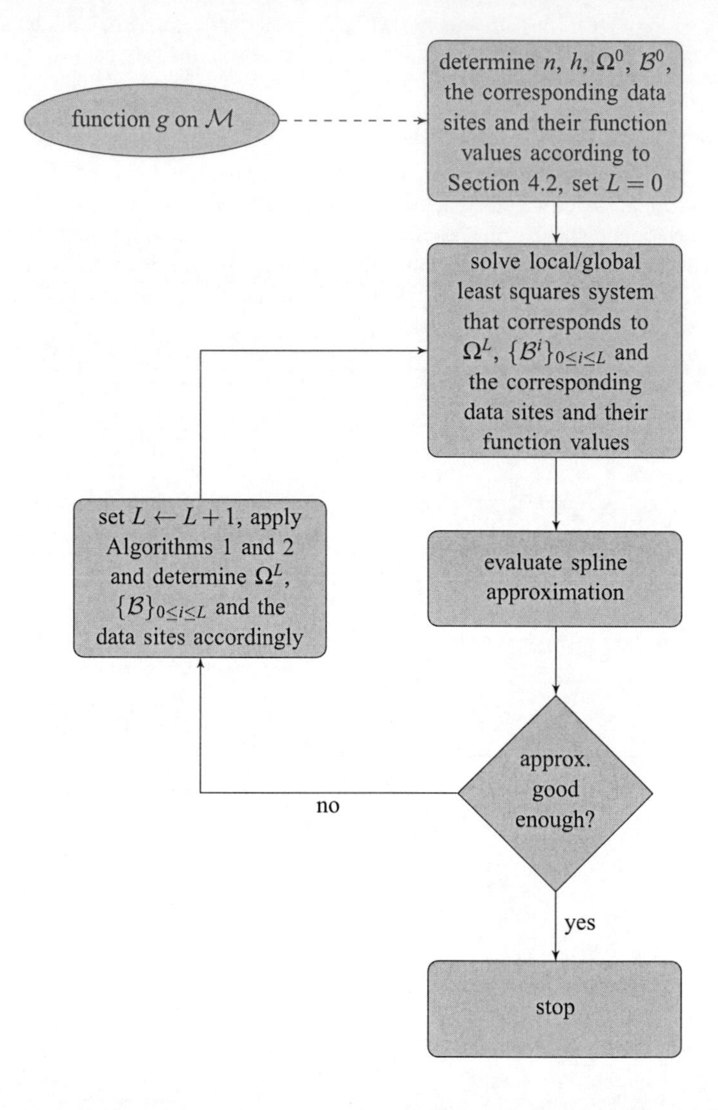

Figure 5.7: Simplified flow chart of the ambient B-spline method with hierarchical refinement.

6 Revised Examples Using Adaptive Refinement

We present the approximation results for the examples already examined in Section 4.3 but apply the hierarchical approaches introduced in Chapter 5. The geoid is treated with the heuristic method of Section 5.2 because of its irregular structure, i.e. there are no connected and distinct regions where hierarchical refinement would be particular meaningful. Contrary, the examples of the Stanford Bunny and the model of the cow are computed with hierarchical B-splines introduced in Section 5.3.

6.1 Geoid

We improve the approximation results of the geoid with the help of the heuristic strategy of hierarchical overlays introduced in Section 5.2. In order to have a closer look at the error behavior and for the sake of reducing computational effort we focus on the region around South America which corresponds to approximately $\frac{1}{12}$ of the earth's surface. The western border of this continent roughly coincides with the course of the Andes and therefore also with high geoid undulations. Figure 6.1 illustrates the approximation error of the spline already discussed before for the region of interest. Of course any other part of the world would have worked, too. As already pointed out when introducing the geoid, our approach is completely local and in order to obtain a globally consistent geoid approximation, we may solve local problems and put them together.

Our goal is to decrease the error at the points of the 15-minute grid within the region of interest below ± 1m. For this we identify all points which violate the chosen threshold. First, we determine the active cells they lie in for the bisected grid size h^1 and compute the residual function values at these points according to (5.2). The error plot for the result of two overlay hierarchies is shown in Figure 6.2, where we not only note a decrease of the error but also of its maximal peaks in the vicinity of the Andes. In order to reach our self-imposed error tolerance we alltogether need five overlay levels. Figures 6.2 - 6.5 reveal the development of the error plot with increasing levels and indicate the used spline coefficients for the corresponding approximation. This example does not confirm the theoretical error behavior but we notice that the relative error drop between level 2 and 3 is significantly greater than before. We have to keep in mind that the hierarchical

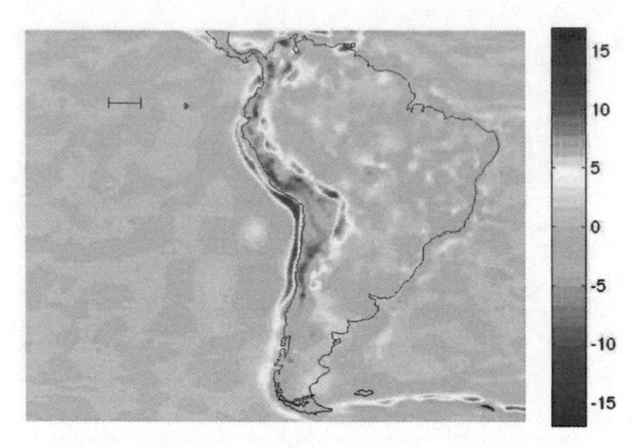

Figure 6.1: Error plot for the spline approximant s^0 with $n = 4$ and grid size $h^0 = 0.1$, where ⊢⊣ illustrates the magnitude of the used grid size. The spline approximant for this part consists of about 1600 coefficients.

overlays are only a heuristic because we only apply the least-squares fit to the residuum remaining. Therefore this example is not a contradiction to the theoretical result.

Figure 6.6 illustrates the used active cells in $\mathbb{R}^3$ for the domain of interest of the different overlay levels. Unsurprisingly, we observe the overlays of the highest level along the course of the Andes where the geoid heights become maximal.

Summing up, compared to the data of the EGM96, our approach overall needs more degrees of freedom to cope with the prescribed error tolerance for the whole geoid model. This is due to the extra dimension we use for the approximation as well as to the choice of the heuristic method which hazards the usage of unnecessary many coefficients. Nevertheless, using the EGM96 always requires about $130,000$ coefficients and spherical harmonics evaluations whereas our approach maximally needs $(L+1)n^3$ B-splines if we use hierachical overlays up to level L. The local supports of the B-splines allow to divide the global approximation into small and manageable problems which can be consistently put together and easily adapted if data is updated locally.

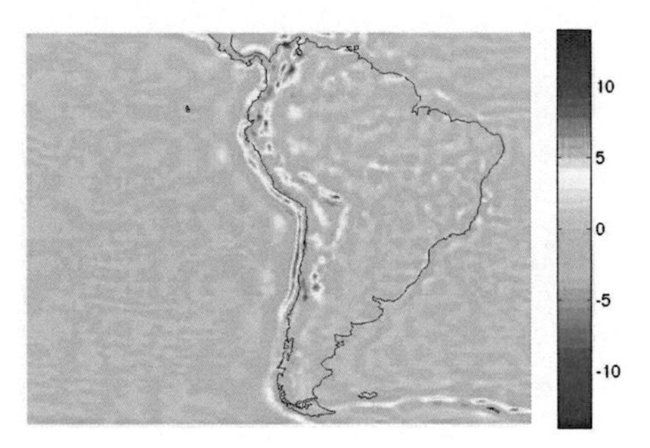

Figure 6.2: Error plot for the spline approximant s^1 with $n = 4$ and grid sizes $h^0 = 0.1$ and $h^1 = 0.05$. The spline approximant for this part consists of about 7000 coefficients.

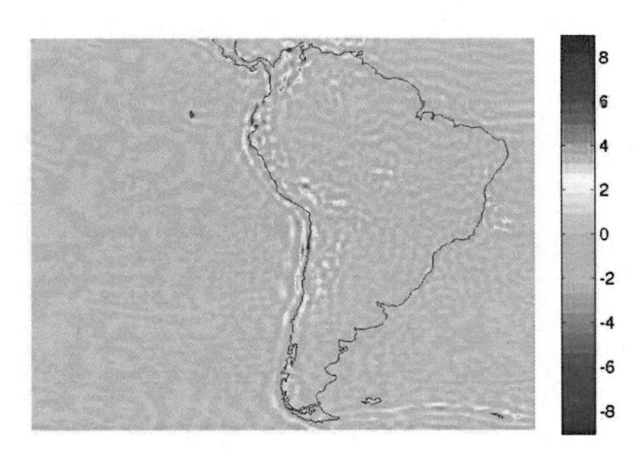

Figure 6.3: Error plot for the spline approximant s^2 with $n = 4$ and grid sizes $h^0 = 0.1$, $h^1 = 0.05$ and $h^2 = 0.025$. The spline approximant for this part consists of about 21000 coefficients.

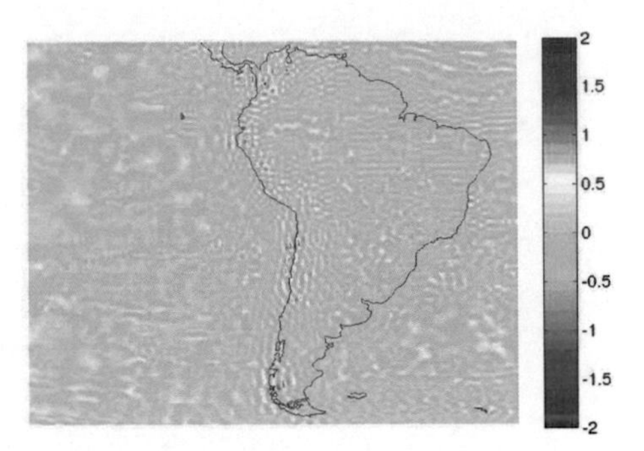

Figure 6.4: Error plot for the spline approximant s^3 with $n = 4$ and grid sizes $h^0 = 0.1$, $h^1 = 0.05$, $h^2 = 0.025$ and $h^3 = 0.0125$. The spline approximant for this part consists of about 56000 coefficients.

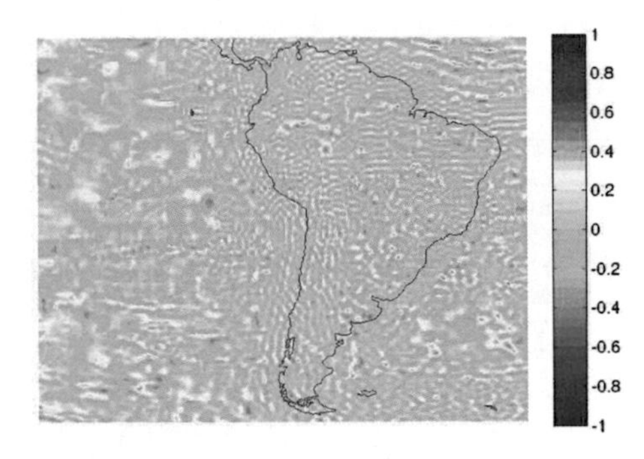

Figure 6.5: Error plot for the spline approximant s^4 with $n = 4$ and grid sizes $h^0 = 0.1$, $h^1 = 0.05$, $h^2 = 0.025$, $h^3 = 0.0125$ and $h^4 = 0.00625$. The spline approximant for this part consists of about 74000 coefficients. The maximal absolute error is less than 0.5m on the 15-minute grid.

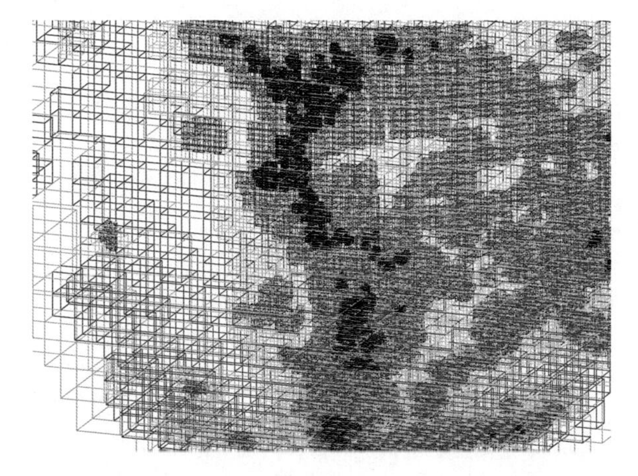

Figure 6.6: The following legend schematically correlates the colors of the active cells to their corresponding overlay level:
$h^0 = \frac{1}{10}$ $h^1 = \frac{1}{20}$ $h^2 = \frac{1}{40}$ $h^3 = \frac{1}{80}$ $h^4 = \frac{1}{160}$.

6.2 Stanford Bunny

In Section 4.3 we noticed the undesirable approximation results around the bunny's neck and ears which were due to the chosen coarse grid size for these areas. The assignment of the projected data sites onto the sphere is done by a linear heuristic. It is therefore not reasonable to analyze the error convergence rates provided by this example as we already mentioned earlier. However, we seek an approximation which acceptably models the vertices of the bunny. We want to use roughly as many B-spline coefficients as the model contains vertices and investigate how close the obtained spline approximant fits the bunny's vertices. The threshold for the refinement rule of the calculated example of the Stanford Bunny was set to $6 \cdot 10^{-3}$.

The approximation starts again with a chosen grid size of $h^0 = 0.1$ and order $n = 4$. The image of the spline approximant of level 0 is illustrated in Figure 4.19 of Section 4.3. The following two iterations are shown in Figures 6.7 and 6.8 where the approximation is computed by two and three hierarchical levels, respectively. Figure 6.9 shows that the bunny calculated by three hierarchical levels mainly consists of B-splines which belong to level 1 (red), i.e. $h^1 = 0.05$. The B-splines of level 2 (cyan) mainly treat the region of the ears and some other difficult parts.

The Figures 6.7 and 6.8 still indicate existing deviations from the original although they comply to the specified error threshold. Nevertheless, when directly comparing the original mesh and the image of the mesh from the sphere, the approximation seems promising, cf. Figure 6.10. For instance, one clearly recognizes the fur structure of the model. We generate a C^2-bunny and may easily obtain any C^{n-2}-bunny by increasing the spline order n.

Furthermore our method can be used for a reparametrization of the polygonal mesh, i.e. any triangulation or quad mesh on the sphere can be transformed into a corresponding parameterization on the bunny. However, due to inhomogeneous distortion factors of the spline approximant, we usually do not obtain a uniform quad mesh on the image. Figures 6.11 and 6.12 illustrate two example quad meshes with different coarseness. We notice that the used speherical parametrization does not tremendously introduce shape distortion in the image of the approximant which is due to the way features of the target surface are mapped onto the reference manifold, such as the bunny's ears. In general, a good reparametrization should respect the underlying distortion in order to produce reasonable results.

6.3 Cow model

We also readdress the cow model of Section 4.3. The iteration starts with the parameters which are already chosen in Figure 4.23, i.e. $h^0 = 0.08$ and $n = 4$. It turns out that even

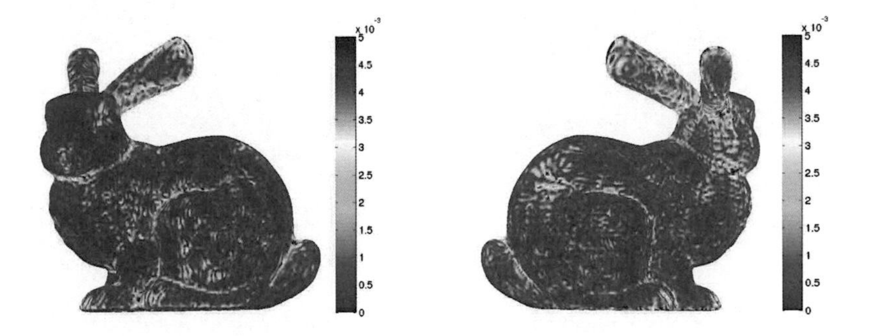

Figure 6.7: This bunny is calculated with 28756 B-spline coefficients and two hierarchy levels. The larger part of the sphere is refined which indicates that the choice for h^0 was not satisfactory. Compared to Figure 4.19 in Subsection 4.3.2, we notice a distinct decrease in the error.

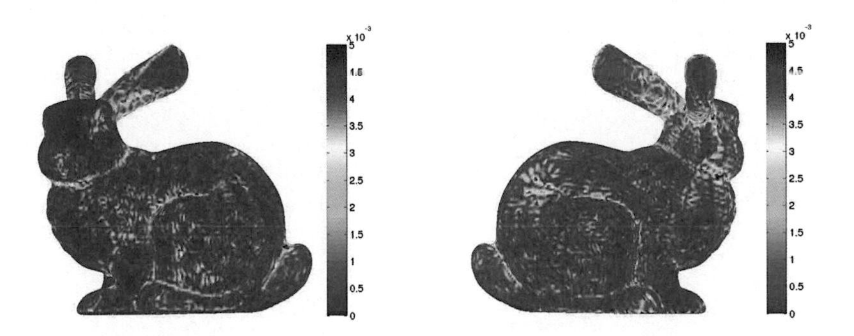

Figure 6.8: The last iteration consists of 37137 B-splines, i.e. roughly about as many as the number of vertices. The root mean squares error is approximately $9 \cdot 10^{-4}$ and the maximal least-squares error is about $6 \cdot 10^{-3}$.

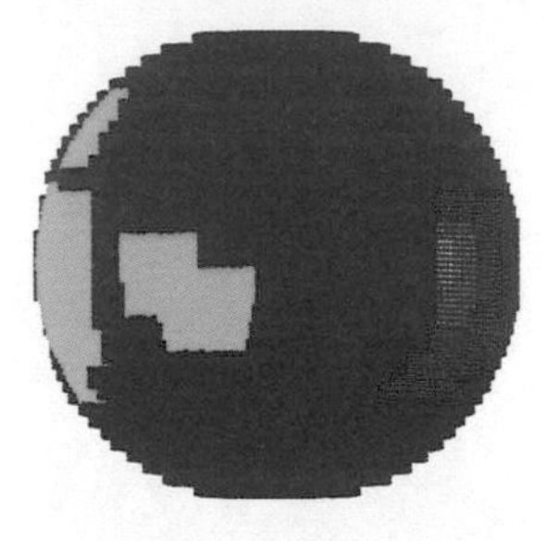

Figure 6.9: The domain for the least squares approximation is colored according to the level the data sites belong to: blue data sites belong to level 0, red data sites to level 1 and cyan data sites to level 2.

the first iteration already consists of about approximately as many degrees of freedom as vertices but clearly fails to represent the surface details. Whereas the cow's body is already modeled well, the hierarchical approach concentrates on the legs, the head, the tail and the udder.

A reasonable result can be seen in Figure 6.13. The obtained spline approximant consists of four hierarchy levels with 86962 B-spline coefficients. The distribution of the levels over the sphere is illustrated in Figure 6.14. We notice that levels 2 and 3 only treat the regions of difficult details. The maximal least squares error $\varepsilon_{\max} \approx 0.03$ and the root mean squares error $\varepsilon_{\text{rms}} \approx 0.003$ from the approximation with one hierarchy level drop to $\varepsilon_{\max} \approx 0.004$ and $\varepsilon_{\text{rms}} \approx 0.0004$ for the result in Figure 6.13.

We have to be aware of the fact that the least-squares errors also depend on the method we determine the function values of the data sites' footpoints on the reference manifold. Especially our linear approximation over the surface triangles reaches its limit if grid sizes become small.

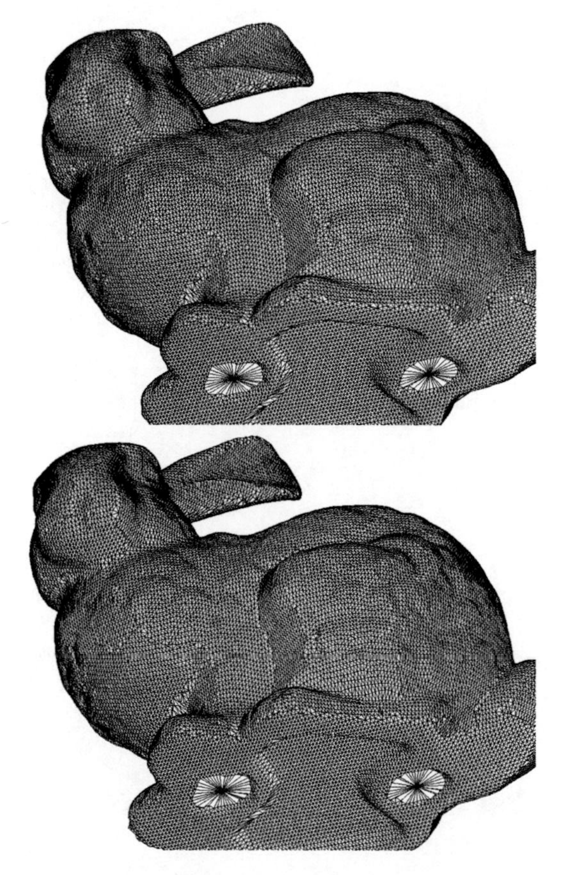

Figure 6.10: Comparison of the image of the spline approximant and the original vertices with its triangulation: at the top we see the approximation with three hierarchy levels, at the bottom the original model data. Although Figure 6.8 indicates that there are deviations, it is not obvious to distinguish the two meshes.

Figure 6.11: Quad mesh on the sphere which is mapped by the spline with three hierarchy levels onto the approximated Stanford Bunny.

Figure 6.12: Very coarse quad mesh on the sphere which is mapped by the spline with three hierarchy levels onto the approximated Stanford Bunny.

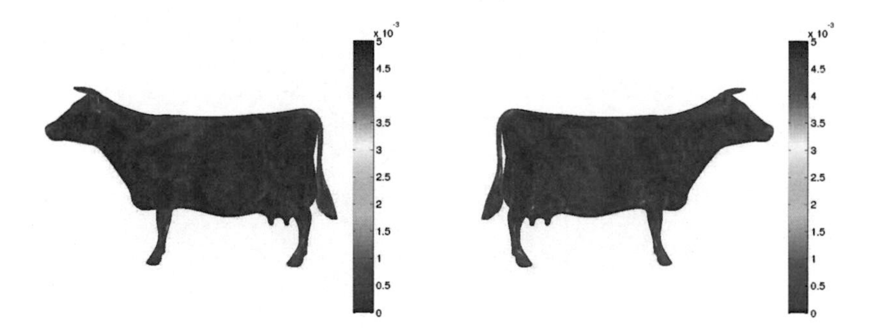

Figure 6.13: The last iteration consists of 86962 B-splines. The root mean squares error is approximately $4 \cdot 10^{-4}$ and the maximal least squares error is about $4 \cdot 10^{-3}$.

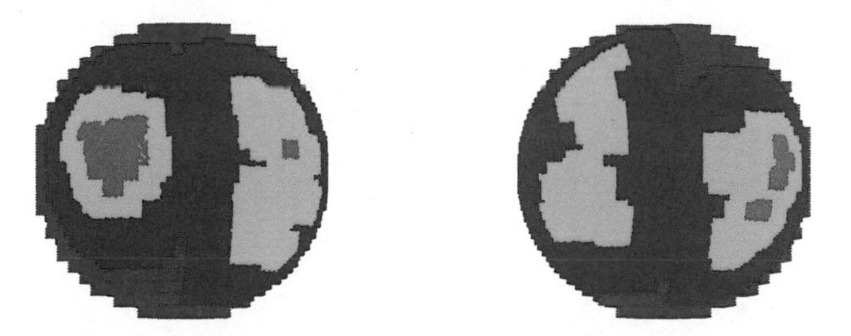

Figure 6.14: The domain for the least squares approximation is colored according to the level the data sites belong to: blue data sites belong to level 0, red data sites to level 1, cyan data sites to level 2 and magenta data sites to level 3.

7 Curvature Analysis

Whereas Chapter 6 shows advantages of the ambient B-spline method in applications, in the following, we want to attend to the analysis of differential geometric quantities such as normal vectors or principal curvatures. Since we technically construct a parametrization, together with a composition of the charts of the reference manifold, one can simply compute the desired quantities according to the classic differential geometry theory. However, we suggest formulas which may generate curvature quantities of the target surface directly from curvature information of the reference manifold.

The Sections 7.1 - 7.3 recapitulate work published in [LR12] except for the second part of Subsection 7.3.3 which additionally provides a numerically robust formula for the ambient B-spline method introduced in this thesis.

7.1 Introduction

The *curvature tensor*, introduced to the field of Applied Geometry by Taubin [Tau95], is commanding increasing interest, presumably due to the following reasons: First, unlike fundamental forms or the shape operator, the curvature tensor is independent of a given parametrization. In particular, this is useful when dealing with piecewise defined surfaces, like spline surfaces, faceted surfaces, or surfaces generated by subdivision. For instance, curvature continuity of a surface is equivalent to continuity of the curvature tensor. This insight was used in [Rei07] to characterize curvature continuity of subdivision schemes. It should be noted that a similar result cannot be formulated in terms of fundamental forms or principal curvatures and -directions. Second, the curvature tensor is defined on the surface itself and thus can be computed equally for surfaces given by a parametrization, in implicit form, or in any other way.

In the following, we present formulas for the computation of the curvature tensor of hypersurfaces in $\mathbb{R}^d$ and discuss also some other aspects like integrability conditions. Most results do not appear in the literature, while a few others are well-known, and included here for the sake of completeness.

After introducing the concept in the next section, we elaborate on different setups: First, we consider parametrized surfaces. Here, the formulas bear a striking resemblance to the standard shape operator, however, without sharing its dependence on the parametrization. In particular, Equation (7.7) suggests an elegant and easy-to-implement scheme for estimating curvature properties of faceted surfaces. Further, we present integrability conditions for reconstructing surfaces from curvature information. These are of substantially lower complexity than when expressed in fundamental forms. Second, we derive formulas for surfaces in implicit form. Amazingly, in the literature, this case is rarely discussed in full detail. For instance, Goldman's nice treatise [Gol05] yields a host of different formulas for the principal (and other) curvatures, but does not address principal directions. Third, we discuss surfaces obtained from a given surface via space deformation. Our results show in detail, how curvature data of the given surface interact with the deformation function to yield the curvature tensor of the new surface.

7.2 The generalized curvature tensor

We start with introducing notations and recalling some basic facts from elementary differential geometry. Since our analysis is local, we consider hypersurfaces in $\mathbb{R}^d$ parametrized by a single function $\mathbf{f}$. More precisely, letting $m := d-1$ throughout, $\mathcal{H}^m$ denotes the space of all m-dimensional hypersurfaces $\mathbf{H} = \mathbf{f}(U) \subset \mathbb{R}^d$, where

- $U \subset \mathbb{R}^m$ is an open domain,
- $\mathbf{f} : U \to \mathbb{R}^d$ is twice differentiable, injective, and regular in the sense that $\operatorname{rank}(\mathrm{d}\mathbf{f}) = m$.

For this chapter, bold face letters are used for objects or functions with values in the geometry space $\mathbb{R}^d$. Vectors are understood as columns, while the differential operator $\mathrm{d} := [\partial_1, \ldots, \partial_m]$ is generating rows. E.g., the Jacobian $\mathrm{d}\mathbf{f} = [\partial_1\mathbf{f}, \ldots, \partial_m\mathbf{f}]$ of $\mathbf{f}$ is a full rank $(d \times m)$-matrix.

The *normal vector field* of $\mathbf{H} \in \mathcal{H}^m$ is a continuous mapping $\mathbf{n} : \mathbf{H} \to \mathbb{S}^m$ from the surface to the m-dimensional unit sphere. The *Gauss map* $\overline{\mathbf{n}} : U \to \mathbb{S}^m$ corresponding to $\mathbf{f}$ is defined by $\overline{\mathbf{n}} := \mathbf{n} \circ \mathbf{f}$ and satisfies

$$\overline{\mathbf{n}}^{\mathrm{t}}\mathrm{d}\mathbf{f} = 0.$$

Differentiating $\overline{\mathbf{n}}^{\mathrm{t}}\overline{\mathbf{n}} = 1$ yields $\overline{\mathbf{n}}^{\mathrm{t}}\mathrm{d}\overline{\mathbf{n}} = 0$, implying $\operatorname{range}\mathrm{d}\overline{\mathbf{n}} \subseteq \operatorname{range}\mathrm{d}\mathbf{f}$. Hence, there is a unique linear map $W : \mathbb{R}^m \to \mathbb{R}^m$, called the *Weingarten map* or also the *shape operator*, satisfying

$$-\mathrm{d}\overline{\mathbf{n}} = \mathrm{d}\mathbf{f}W. \tag{7.1}$$

Multiplication by $\mathrm{d}\mathbf{f}$ from the left yields the normal equation

$$I\!I = IW,$$

where $I := \mathbf{df}^t \mathbf{df}$ and $I\!I := -\mathbf{df}^t \mathbf{d}\bar{\mathbf{n}}$ are the first and second fundamental form of $\mathbf{f}$, respectively. By regularity of $\mathbf{f}$, the matrix I is invertible. Hence, we may solve for W to find

$$W = I^{-1} I\!I = -\mathbf{df}^+ \mathbf{d}\bar{\mathbf{n}}, \tag{7.2}$$

where

$$\mathbf{df}^+ := I^{-1} \mathbf{df}^t \tag{7.3}$$

is the pseudo-inverse of $\mathbf{df}$. Further analysis reveals that W is self-adjoint with respect to I. This fact guarantees existence of a complete set of real eigenvalues κ_i, called the *principal curvatures* of $\mathbf{H}$, corresponding to eigenspaces V_i,

$$V_i := \{v \in \mathbb{R}^m : Wv = \kappa_i v\}, \quad i = 1, \ldots, m.$$

It is easy to see that $v \in V_i$ implies

$$\mathbf{d}\bar{\mathbf{n}}\, v = -\kappa_i \mathbf{df} v. \tag{7.4}$$

When mapping vectors $v \in V_i$ from the parameter space $\mathbb{R}^m$ to geometry space $\mathbb{R}^d$, we obtain *principal directions* $\mathbf{v} := \mathbf{df}\, v$ corresponding to κ_i. The according *principal subspaces* are denoted by $\mathbf{V}_i := \{\mathbf{df}\, v : v \in V_i\}$.

It is well known that the principal curvatures κ_i and subspaces $\mathbf{V}_i$ are uniquely defined (up to ordering) and independent of the given parametrization of the surface $\mathbf{H}$. By contrast, the matrices $I, I\!I, W$ depend on the given parametrization. When considering piecewise defined surfaces, as generated by spline or subdivision techniques, this fact may result in discontinuities of these objects at patch boundaries, even if the surface is geometrically smooth. Examples can be found in [PR08]. The concept of curvature tensors resolves this problem. We characterize it as a linear map in $\mathbb{R}^d$ with κ_i and $\mathbf{v} \in \mathbf{V}_i$ as eigenvalues and -vectors, respectively. The normal vector $\mathbf{n}$ is a further eigenvector, where the corresponding eigenvalue μ can be prescribed at will. Of course, the canonical choice is $\mu = 0$, but for instance, the case that $\mu = -\kappa_{\mathrm{m}}$ equals the negative mean curvature is considered in [HP11]. Below,

$$\mathcal{E}^d := \{\mathbf{A} \in \mathbb{R}^{d \times d} : \mathbf{A}^t = \mathbf{A}\}$$

denotes the space of symmetric endomorphisms in $\mathbb{R}^d$.

Definition 7.1. *Let $\mathbf{H} \in \mathcal{H}^m$ be a hypersurface with normal vector field $\mathbf{n}$, principal curvatures $\kappa_1, \ldots, \kappa_{d-1}$, and principal subspaces $\mathbf{V}_1, \ldots, \mathbf{V}_{d-1}$. Further, let $\mu : \mathbf{H} \to \mathbb{R}$ be some function defined on $\mathbf{H}$. The mapping $\mathbf{E}_\mu : \mathbf{H} \to \mathcal{E}^d$, characterized by*

$$\mathbf{E}_\mu \mathbf{n} = \mu \mathbf{n} \quad \textit{and} \quad \mathbf{E}_\mu \mathbf{v} = \kappa_i \mathbf{v}, \quad \mathbf{v} \in \mathbf{V}_i,\ i = 1, \ldots, d-1,$$

is called the generalized curvature tensor of $\mathbf{H}$ corresponding to μ. For $\mu = 0$, we obtain the special case $\mathbf{E} := \mathbf{E}_0$, which is simply called the curvature tensor or also the embedded Weingarten map of $\mathbf{H}$.

It is important to note that, by definition, curvature tensors do not depend on the parametrization, but only on the geometry of $\mathbf{H}$ and the choice of μ. Clearly, the mean curvature κ_{m} and the Gaussian curvature κ_{g} are given by

$$\kappa_{\mathrm{m}} = \frac{\operatorname{trace}\mathbf{E}_\mu - \mu}{m}, \qquad \kappa_{\mathrm{g}} = \frac{\det \mathbf{E}_\mu}{\mu},$$

respectively, the latter of course only if $\mu \neq 0$.

Finally, we note that $\mathbf{E}$ and $\mathbf{E}_\mu$ are related by

$$\mathbf{E}_\mu = \mathbf{E} + \mu \mathbf{n}\mathbf{n}^{\mathrm{t}}.$$

To show this, we consider two cases: First, $\mathbf{E}_\mu \mathbf{n} = \mu\mathbf{n} = \mathbf{E}\mathbf{n} + \mu\mathbf{n}\mathbf{n}^{\mathrm{t}}\mathbf{n}$ since $\mathbf{E}\mathbf{n} = 0$ and $\mathbf{n}^{\mathrm{t}}\mathbf{n} = 1$. Second, if $\mathbf{v} = \mathrm{d}\mathbf{f}\, v$ is a principal direction with associated principal curvature κ, then $\mathbf{n}^{\mathrm{t}}\mathbf{v} = 0$ so that $\mathbf{E}_\mu\mathbf{v} = \mathbf{E}\mathbf{v} = \kappa\mathbf{v}$, as requested. As a consequence of the latter display, it is sufficient to specify formulas for $\mathbf{E}$.

7.3 Three different setups

In the following subsections, we consider the curvature tensor of hypersurfaces given in three different forms: parametrized, implicit, and resulting from space deformation.

7.3.1 Parametrized surfaces

Formulas for the curvature tensor of parametrized surfaces $\mathbf{H} = \mathbf{f}(U)$, given in Theorem 7.2, strongly resemble those for the standard shape operator. As briefly sketched then, the crucial relation (7.7) suggests a most simple scheme for estimating curvature properties of faceted surfaces. Further, in Theorem 7.3, we provide integrability conditions which admit the reconstruction of a surface from given curvature data. The case of parametrized surfaces in $\mathbb{R}^3$ has been discussed in a similar way in [Rei07].

Analogous to the Gauss map $\bar{\mathbf{n}} := \mathbf{n} \circ \mathbf{f} : U \to \mathbb{S}^m$, we introduce the function $\bar{\mathbf{E}} := \mathbf{E} \circ \mathbf{f} : U \to \mathcal{E}^d$, and call it the *parametrized curvature tensor*. By Definition 7.1, the conditions characterizing $\bar{\mathbf{E}}$ are

$$\bar{\mathbf{E}}\bar{\mathbf{n}} = 0 \quad \text{and} \quad \bar{\mathbf{E}}\mathbf{v} = \kappa_i \mathbf{v}, \quad \mathbf{v} \in \mathbf{V}_i,\ i = 1, \ldots, d-1.$$

In this setting, everything depends on the argument $u \in U$, which is omitted throughout to improve readability. The following theorem provides explicit formulas for the computation of $\bar{\mathbf{E}}$:

Theorem 7.2. *Let* $\mathbf{H} = \mathbf{f}(U) \in \mathcal{H}^m$ *be a hypersurface with Gauss map* $\bar{\mathbf{n}}$*. The parametrized curvature tensor* $\bar{\mathbf{E}}$ *is given by*

$$\bar{\mathbf{E}} = -\mathrm{d}\bar{\mathbf{n}}\,\mathrm{d}\mathbf{f}^{+} \tag{7.5}$$

or, equivalently, by

$$\bar{\mathbf{E}} = (\mathrm{d}\mathbf{f}^{+})^{\mathrm{t}}\,I\!I\,\mathrm{d}\mathbf{f}^{+}, \tag{7.6}$$

where $\mathrm{d}\mathbf{f}^{+} = I^{-1}\mathrm{d}\mathbf{f}^{\mathrm{t}}$ *is the pseudo-inverse of* $\mathrm{d}\mathbf{f}$*, as defined in* (7.3)*, and* $I, I\!I$ *are the first and second fundamental form of* $\mathbf{f}$*. Furthermore,*

$$-\mathrm{d}\bar{\mathbf{n}} = \bar{\mathbf{E}}\,\mathrm{d}\mathbf{f}. \tag{7.7}$$

Proof: We denote the matrices given by the right hand sides of (7.5) and (7.6) by $\bar{\mathbf{E}}_1$ and $\bar{\mathbf{E}}_2$, respectively. First, $\mathrm{d}\mathbf{f}^{\mathrm{t}}\,\bar{\mathbf{n}} = 0$ shows that $\bar{\mathbf{E}}_1\bar{\mathbf{n}} = \bar{\mathbf{E}}_2\bar{\mathbf{n}} = 0$. Now, let $\mathbf{v} = \mathrm{d}\mathbf{f}\upsilon \in \mathbf{V}_i$ be a principal direction corresponding to κ_i. We obtain using (7.4)

$$\bar{\mathbf{E}}_1\mathbf{v} = -\mathrm{d}\bar{\mathbf{n}}I^{-1}\,\mathrm{d}\mathbf{f}^{\mathrm{t}}\mathrm{d}\mathbf{f}\,\upsilon = -\mathrm{d}\bar{\mathbf{n}}\,\upsilon = \kappa_i\mathrm{d}\mathbf{f}\,\upsilon = \kappa_i\mathbf{v},$$

and equally, using $I\!I = -\mathrm{d}\mathbf{f}^{\mathrm{t}}\,\mathrm{d}\bar{\mathbf{n}}$,

$$\bar{\mathbf{E}}_2\mathbf{v} = \mathrm{d}\mathbf{f}I^{-1}\,\mathrm{d}\mathbf{f}^{\mathrm{t}}\,\bar{\mathbf{E}}_1\,\mathbf{v} = \kappa_i\mathrm{d}\mathbf{f}I^{-1}\,\mathrm{d}\mathbf{f}^{\mathrm{t}}\mathrm{d}\mathbf{f}\,\upsilon = \kappa_i\mathrm{d}\mathbf{f}\,\upsilon = \kappa_i\mathbf{v}.$$

Hence, $\bar{\mathbf{E}} = \bar{\mathbf{E}}_1 = \bar{\mathbf{E}}_2$. Finally, multiplying (7.5) by $\mathrm{d}\mathbf{f}$ from the right yields (7.7). □

Let us briefly comment on the three equations appearing in the theorem: First, when comparing (7.5) with (7.2), we find that W and $\bar{\mathbf{E}}$ differ just by the ordering of their factors,

$$W = -\mathrm{d}\mathbf{f}^{+}\,\mathrm{d}\bar{\mathbf{n}}, \quad \bar{\mathbf{E}} = -\mathrm{d}\bar{\mathbf{n}}\,\mathrm{d}\mathbf{f}^{+}\,.$$

Second, since the second fundamental form $I\!I$ is symmetric, (7.6) shows that also $\bar{\mathbf{E}}$ is symmetric, thus reconfirming the well known fact that the principal curvatures are all real and that it is possible to choose an orthonormal set of principal directions $\mathbf{v}_1, \ldots, \mathbf{v}_m$. Third, when comparing (7.7) with (7.1), we see that both operators establish a relation between $\mathrm{d}\bar{\mathbf{n}}$ and $\mathrm{d}\mathbf{f}$,

$$-\mathrm{d}\bar{\mathbf{n}} = \mathrm{d}\mathbf{f}W, \quad -\mathrm{d}\bar{\mathbf{n}} = \bar{\mathbf{E}}\,\mathrm{d}\mathbf{f}.$$

However, W is acting on the parameter space $\mathbb{R}^m$, while $\bar{\mathbf{E}}$ is acting on the geometry space $\mathbb{R}^d$, in which the surface $\mathbf{H}$ is embedded. This observation accounts for the name *embedded Weingarten map*, as it was proposed in [Rei07].

There exist a few schemes for estimating the curvature tensor on faceted surfaces, see e.g. [TRZS04, Gol04, Tau95, CSM03]. Compared with these approaches, (7.7) suggests an extremely simple alternative: Given vertices $\mathbf{p}_j \in \mathbf{H}$ and (possibly estimated) corresponding normal vectors $\mathbf{n}_j \in \mathbb{S}^m, j = 1,\ldots,J$, one can proceed as follows. Without loss of generality, let us consider some triangle $\mathbf{H}_\Delta$ with vertices $\mathbf{p}_1, \mathbf{p}_2, \mathbf{p}_3$, approximating some part $\mathbf{H}_0$ of the surface $\mathbf{H}$. An approximation $\mathbf{E}_\Delta$ of the curvature tensor $\mathbf{E}_0$ at the center of $\mathbf{H}_0$ is sought. Regarding $\mathbf{H}_\Delta = \mathbf{f}_\Delta(T)$ and $\mathbf{H}_0 = \mathbf{f}_0(T)$ as images of the unit triangle T, we may employ the trivial discretization

$$\mathrm{d}\mathbf{f}_0 \approx \mathrm{d}\mathbf{f}_\Delta := [\mathbf{p}_2 - \mathbf{p}_1,\, \mathbf{p}_3 - \mathbf{p}_1], \quad \mathrm{d}\bar{\mathbf{n}}_0 \approx \mathrm{d}\bar{\mathbf{n}}_\Delta := [\mathbf{n}_2 - \mathbf{n}_1,\, \mathbf{n}_3 - \mathbf{n}_1]$$

to obtain the matrix

$$\tilde{\mathbf{E}}_\Delta := -\mathrm{d}\bar{\mathbf{n}}_0\, \mathrm{d}\mathbf{f}_\Delta^+,$$

which is already an approximation of $\mathbf{E}_0$. However, to remedy the typical lack of symmetry of $\tilde{\mathbf{E}}_\Delta$, we define

$$\mathbf{E}_\Delta := (\mathrm{Id} - \mathbf{n}_\Delta \mathbf{n}_\Delta^{\mathrm{t}})(\tilde{\mathbf{E}}_\Delta + \tilde{\mathbf{E}}_\Delta^{\mathrm{t}})(\mathrm{Id} - \mathbf{n}_\Delta \mathbf{n}_\Delta^{\mathrm{t}}), \quad \mathbf{n}_\Delta := (\mathbf{n}_1 + \mathbf{n}_2 + \mathbf{n}_3)/3.$$

One can show that this matrix is closest to $\tilde{\mathbf{E}}_\Delta$ with respect to the Frobenius norm among all symmetric matrices satisfying $\mathbf{E}_\Delta \mathbf{n}_\Delta = 0$ for given $\mathbf{n}_\Delta$. First experiments [Leh09] indicate that this approach combines extremely simple implementation with approximation order $\|\mathbf{E}_0 - \mathbf{E}_\Delta\| = O(h^2)$ provided that sufficiently accurate normal vectors are available.

Another interesting application of curvature tensors concerns the reconstruction of surfaces from given curvature information. When prescribing fundamental forms, fairly complicated integrability conditions, known as Gauss and Mainardi-Codazzi equations, have to be obeyed. By contrast, the integrability conditions on the parametrized curvature tensor have a *much* simpler form.

Theorem 7.3. *Let $U \subset \mathbb{R}^m$ be an open, simply connected domain. Consider any C^2-function $\bar{\mathbf{E}} : U \to \mathcal{E}^d$ with the following properties:*

- *$\bar{\mathbf{E}}$ has a simple eigenvalue $\mu = 0$. The corresponding normalized eigenvector is denoted by $\bar{\mathbf{n}} : U \to \mathbb{S}^m$, where the sign is chosen consistently to obtain a continuous map.*

- *For all $k, \ell = 1,\ldots,m$, it holds*

$$\partial_k \bar{\mathbf{E}}^+ \partial_\ell \bar{\mathbf{n}} = \partial_\ell \bar{\mathbf{E}}^+ \partial_k \bar{\mathbf{n}}, \tag{7.8}$$

where $\bar{\mathbf{E}}^+$ denotes the pseudo-inverse of $\bar{\mathbf{E}}$.

Then there exists a parametrization $\mathbf{f} : U \to \mathbb{R}^d$, *unique up to translation, such that* $\overline{\mathbf{n}}$ *and* $\overline{\mathbf{E}}$ *are the Gauss map and the parametrized curvature tensor of* $\mathbf{f}$, *respectively. Further, no such parametrization exists if the second property above is violated.*

Proof: We recall that, by definition of the pseudo-inverse, $\overline{\mathbf{E}}^+\overline{\mathbf{n}} = 0$, and $\overline{\mathbf{E}}^+\mathbf{v} = \kappa_i^{-1}\mathbf{v}$ for any $\mathbf{v} \in \mathbf{V}_i$. In particular, $\overline{\mathbf{E}}^+ \in \mathcal{E}^d$ is symmetric, too. Let us assume that $\overline{\mathbf{E}}$ and $\overline{\mathbf{n}}$ are given as specified above. Since 0 is a simple eigenvalue of $\overline{\mathbf{E}}$, the vector field $\overline{\mathbf{n}}$ satisfying $\overline{\mathbf{E}}\overline{\mathbf{n}} = 0$ is not only continuous, but as smooth as $\overline{\mathbf{E}}$. Hence, $\partial_k\partial_\ell\overline{\mathbf{n}} = \partial_\ell\partial_k\overline{\mathbf{n}}$. Define the $(d \times m)$-matrix $\mathbf{J} := -\overline{\mathbf{E}}^+\mathrm{d}\overline{\mathbf{n}}$, and denote its columns by $\mathbf{J}_1, \ldots, \mathbf{J}_m$. Then, by (7.8),

$$\partial_k\mathbf{J}_\ell = -\partial_k(\overline{\mathbf{E}}^+\partial_\ell\overline{\mathbf{n}}) = -\partial_k\overline{\mathbf{E}}^+\partial_\ell\overline{\mathbf{n}} - \overline{\mathbf{E}}^+\partial_k\partial_\ell\overline{\mathbf{n}}$$

equals

$$\partial_\ell\mathbf{J}_k = -\partial_\ell(\overline{\mathbf{E}}^+\partial_k\overline{\mathbf{n}}) = -\partial_\ell\overline{\mathbf{E}}^+\partial_k\overline{\mathbf{n}} - \overline{\mathbf{E}}^+\partial_\ell\partial_k\overline{\mathbf{n}}.$$

Hence, there exists a potential $\mathbf{f} : U \to \mathbb{R}^d$ of $\mathbf{J}$, i.e., $\mathrm{d}\mathbf{f} = \mathbf{J}$. The vector field $\overline{\mathbf{n}}$ is the Gauss map of $\mathbf{f}$ since $\overline{\mathbf{n}}^t\mathrm{d}\mathbf{f} = -\overline{\mathbf{n}}^t\overline{\mathbf{E}}^+\mathrm{d}\overline{\mathbf{n}} = 0$ by the first condition on $\overline{\mathbf{E}}$. Further, $\mathrm{d}\mathbf{f} = -\overline{\mathbf{E}}^+\mathrm{d}\overline{\mathbf{n}}$ implies $\overline{\mathbf{E}}\mathrm{d}\mathbf{f} = -\mathrm{d}\overline{\mathbf{n}}$, showing that $\overline{\mathbf{E}}$ is indeed the parametrized curvature tensor of $\mathbf{f}$. Let us assume that $\mathbf{f}'$ is another parametrization complying with $\overline{\mathbf{n}}$ and $\overline{\mathbf{E}}$. Then $\mathrm{d}\mathbf{f} = \mathrm{d}\mathbf{f}' = -\overline{\mathbf{E}}^+\mathrm{d}\overline{\mathbf{n}}$ yields $\mathrm{d}(\mathbf{f} - \mathbf{f}') = 0$, showing that $\mathbf{f}$ and $\mathbf{f}'$ differ only by a constant. To establish necessity of (7.8), let us assume that there exists $\mathbf{f}$ with $\mathrm{d}\mathbf{f} = -\overline{\mathbf{E}}^+\mathrm{d}\overline{\mathbf{n}}$. Then $\mathbf{f}$ is at least as smooth as $\overline{\mathbf{E}}$ so that $\partial_k\partial_\ell\mathbf{f} = \partial_\ell\partial_k\mathbf{f}$ must hold. Now, (7.8) follows immediately. □

7.3.2 Implicit surfaces

Curvature formulas are less commonly stated for hypersurfaces that are level sets of real-valued functions. Only recently, Goldman [Gol05] collected old and provided new formulas for the principal curvatures in this context. Our results addressing the curvature tensor are different and reveal a simple underlying structure. Of course, the result (7.11) on the signed distance function is well known, but to the best of our knowledge, (7.10) is new in this form. The matrix defined in [BPK98] is somehow similar, but it is not symmetric, and the surface normal is a left rather than a right eigenvector.

The C^2-function $F : \Sigma \to \mathbb{R}$, defined on some open neighborhood $\Sigma \subset \mathbb{R}^d$ of $\mathbf{H}$, is called an *implicit representation* of the hypersurface $\mathbf{H} \in \mathcal{H}^m$, if

- $F_{|\mathbf{H}} = 0$, and
- F is regular in the sense that $\|\mathrm{D}F\| \neq 0$.

While d denotes differentiation with respect to parameters in $\mathbb{R}^m$, the operator $\mathrm{D} := [\partial_1, \ldots, \partial_d]$ denotes differentiation with respect to the coordinates in geometry space $\mathbb{R}^d$. It is our goal to establish formulas for the curvature tensor $\mathbf{E}$ in terms of the function F.

Given a parametrization $\mathbf{f} : U \to \mathbb{R}^d$ of $\mathbf{H}$, equality of two functions $\Phi, \tilde{\Phi} : \Sigma \to \mathbb{R}$ on $\mathbf{H}$ can be shown by the following simple argument:

$$\Phi \circ \mathbf{f} = \tilde{\Phi} \circ \mathbf{f} \quad \text{if and only if} \quad \Phi_{|\mathbf{H}} = \tilde{\Phi}_{|\mathbf{H}}.$$

In particular, $F_{|\mathbf{H}} = 0$ is equivalent to $F \circ \mathbf{f} = 0$. Hence, by the chain rule,

$$0 = \mathrm{d}(F \circ \mathbf{f}) = (\mathrm{D}F \circ \mathbf{f})\mathrm{d}\mathbf{f}.$$

Defining the function $\mathbf{N} : \Sigma \to \mathbb{S}^m$ by

$$\mathbf{N} := \frac{\mathrm{D}^{\mathrm{t}}F}{\|\mathrm{D}F\|},$$

this implies that $\mathbf{N} \circ \mathbf{f}$ is perpendicular to $\mathbf{H}$. In other words,

$$\mathbf{n} := -\mathbf{N}_{|\mathbf{H}} \quad \text{and} \quad \bar{\mathbf{n}} = \mathbf{n} \circ \mathbf{f} = -\mathbf{N} \circ \mathbf{f}$$

are the normal vector field of $\mathbf{H}$ and the corresponding Gauss map, respectively. The deliberate choice of the orientation of the normal vector field opposite to $\mathbf{N}$ will yield simpler formulas later on. Below, $\mathrm{D}^2F = (\partial_i \partial_j F)_{i,j}$ denotes the Hessian of F. Further, we use the following convention: If functions defined on Σ and functions defined on the subset $\mathbf{H} \subset \Sigma$ appear in the same formula, then the functions defined on Σ are understood to be restricted to $\mathbf{H}$.

Theorem 7.4. *Let $\mathbf{H} \in \mathcal{H}^m$ be a hypersurface with implicit representation $F : \Sigma \to \mathbb{R}$. With $\mathbf{N}$ as defined above and $\mathbf{T} := \mathrm{Id} - \mathbf{n}\mathbf{n}^{\mathrm{t}} : \mathbf{H} \to \mathcal{E}^d$ the orthogonal projector onto the tangent space, the curvature tensor is given by*

$$\mathbf{E} = \mathrm{D}\mathbf{N}\,\mathbf{T} \tag{7.9}$$

or, equivalently, by

$$\mathbf{E} = \frac{1}{\|\mathrm{D}F\|}\mathbf{T}\,\mathrm{D}^2F\,\mathbf{T}. \tag{7.10}$$

In particular, if F is a signed distance function, i.e., $\|\mathrm{D}F\| = 1$, then

$$\mathbf{E} = \mathrm{D}^2F. \tag{7.11}$$

Proof: We denote the matrices given by the right hand sides of (7.9) and (7.10) by $\mathbf{E}_1$ and $\mathbf{E}_2$, respectively. First, $\mathbf{Tn} = 0$ yields $\mathbf{E}_1\mathbf{n} = \mathbf{E}_2\mathbf{n} = 0$. Now, let $\mathbf{v} = \mathrm{d}\mathbf{f}v \in \mathbf{V}_i$ be a principal direction corresponding to κ_i. We obtain using (7.4) and the chain rule

$$\begin{aligned}(\mathbf{E}_1 \circ \mathbf{f})\mathbf{v} &= (\mathrm{D}\mathbf{N} \circ \mathbf{f})(\mathrm{Id} - \bar{\mathbf{n}}\bar{\mathbf{n}}^{\mathrm{t}})\mathrm{d}\mathbf{f}\,v = (\mathrm{D}\mathbf{N} \circ \mathbf{f})\mathrm{d}\mathbf{f}\,v \\ &= -\mathrm{d}\bar{\mathbf{n}}\,v = \kappa_i \mathrm{d}\mathbf{f}\,v = \kappa_i \mathbf{v}.\end{aligned}$$

Thus, $\mathbf{E} = \mathbf{E}_1$. Regarding (7.10), we differentiate the identity $\|\mathrm{D}F\|\,\mathbf{N} = \mathrm{D}^{\mathrm{t}}F$ using $\mathrm{D}\|\mathrm{D}F\| = (\mathrm{D}F\,\mathrm{D}^2F)/\|\mathrm{D}F\| = \mathbf{N}^{\mathrm{t}}\,\mathrm{D}^2F$. This yields $\mathbf{N}\mathbf{N}^{\mathrm{t}}\,\mathrm{D}^2F + \|\mathrm{D}F\|\,\mathrm{D}\mathbf{N} = \mathrm{D}^2F$. Solving for $\mathrm{D}\mathbf{N}$, we find

$$\mathrm{D}\mathbf{N} = \frac{1}{\|\mathrm{D}F\|}\,\mathbf{T}\,\mathrm{D}^2F$$

and

$$\mathbf{E}_2 = \frac{1}{\|\mathrm{D}F\|}\,\mathbf{T}\,\mathrm{D}^2F\,\mathbf{T} = \mathrm{D}\mathbf{N}\mathbf{T} = \mathbf{E}_1 = \mathbf{E}.$$

Finally, if $\|\mathrm{D}F\| = 1$, we differentiate $\mathrm{D}F\,\mathrm{D}^{\mathrm{t}}F = 1$ to obtain $\mathrm{D}F\,\mathrm{D}^2F = 0$. Hence, by symmetry of the Hessian, $\mathbf{N}^{\mathrm{t}}\,\mathrm{D}^2F = 0$ and $\mathrm{D}^2F\,\mathbf{N} = 0$, showing that $\mathbf{T}\,\mathrm{D}^2F\,\mathbf{T} = \mathrm{D}^2F$. □

7.3.3 Space deformation

The ambient B-spline method may be regarded as defining surfaces from an initial surface by a deformation of the embedding space, as we already discussed. The general idea of space deformation is used, for instance, in the context of surface morphing. Here, we want to relate curvature properties of the deformed surface with curvature properties of the initial surface and both the vector field representing the space deformation and its inverse. We are not aware of any references addressing this issue in a systematic way.

Consider a hypersurface $\mathbf{H} \in \mathcal{H}^m$ with normal vector field $\mathbf{n}$ and curvature tensor $\mathbf{E}$. As before, $\Sigma \subset \mathbb{R}^d$ denotes some open neighborhood of $\mathbf{H}$. Given a C^2-diffeomorphism $\Phi : \Sigma \to \Sigma_* \subset \mathbb{R}^d$, we define the new surface $\mathbf{H}_* := \Phi(\mathbf{H})$. That is, $\mathbf{H}$ is transformed into $\mathbf{H}_*$ by deforming the surrounding geometry space. Denoting the inverse of Φ by $\Phi_* : \Sigma_* \to \Sigma$, we may also write $\mathbf{H} = \Phi_*(\mathbf{H}_*)$. Depending on the application, either Φ or Φ_* may be known, and it is our goal to derive formulas for the curvature tensor $\mathbf{E}_*$ of $\mathbf{H}_*$ in terms of $\mathbf{E}, \mathbf{n}$, and either Φ or Φ_*. The Jacobians of Φ and Φ_* are denoted by

$$\mathbf{J} := \mathrm{D}\Phi \quad \text{and} \quad \mathbf{J}_* := \mathrm{D}\Phi_*,$$

respectively. Clearly, these are inverse matrices,

$$\mathbf{J}\mathbf{J}_* = \mathrm{Id}\,.$$

Here and below, we use the following convention: If functions defined on Σ and Σ_* appear in the same formula, then they are understood to be evaluated at corresponding points $\mathbf{x} \in \Sigma$ and $\mathbf{x}_* \in \Sigma_*$, related by $\mathbf{x} = \Phi_*(\mathbf{x}_*)$ and $\mathbf{x}_* = \Phi(\mathbf{x})$. For instance, the above formula actually reads $\mathbf{J}(\mathbf{x})\mathbf{J}_*(\mathbf{x}_*) = \mathrm{Id}$.

Referring to Subsection 7.3.2, let $F : \Sigma \to \mathbb{R}$ be the signed distance function of $\mathbf{H}$, i.e., $F_{|\mathbf{H}} = 0$ and $\|\mathrm{D}F\| = 1$. Then $F_* := F \circ \Phi_*$ is an implicit representation of $\mathbf{H}_*$. By the chain rule, $\mathrm{D}F_* = \mathrm{D}F\,\mathbf{J}_*$ and $\mathrm{D}F = \mathrm{D}F_*\,\mathbf{J}$. Hence, the unit vector fields $\mathbf{N}$ and $\mathbf{N}_*$ corresponding to F and F_*, are related by

$$\mathbf{N}_* = \frac{\mathbf{J}_*^{\mathrm{t}}\mathbf{N}}{\|\mathbf{J}_*^{\mathrm{t}}\mathbf{N}\|} = \frac{\mathbf{J}^{-\mathrm{t}}\mathbf{N}}{\|\mathbf{J}^{-\mathrm{t}}\mathbf{N}\|}.$$

When restricting $-\mathbf{N}$ and $-\mathbf{N}_*$ to $\mathbf{H}$ and $\mathbf{H}_*$, we obtain the normal vector fields $\mathbf{n}$ and $\mathbf{n}_*$, respectively. Below, we use the notation

$$\mathrm{D}^2_{\mathbf{v}}\Phi := \sum_{i=1}^{d} \mathbf{v}_i \mathrm{D}^2\Phi_i$$

for the linear combination of Hessians of the coordinate functions of Φ with the components of the vector $\mathbf{v} \in \mathbb{R}^d$.

Theorem 7.5. *Let $\mathbf{H} \in \mathcal{H}^m$ be a hypersurface with curvature tensor $\mathbf{E}$, and $\mathbf{H}_* := \Phi(\mathbf{H})$ for some C^2-diffeomorphism, as introduced above. Then, with $\mathbf{T}_* := \mathrm{Id} - \mathbf{n}_*\mathbf{n}_*^{\mathrm{t}}$ and the notation introduced above, the curvature tensor of $\mathbf{H}_*$ is given in terms of Φ_* by*

$$\mathbf{E}_* = \|\mathbf{J}_*^{\mathrm{t}}\mathbf{n}\|^{-1}\,\mathbf{T}_*\,(\mathbf{J}_*^{\mathrm{t}}\,\mathbf{E}\,\mathbf{J}_* - \mathrm{D}^2_{\mathbf{n}}\Phi_*)\,\mathbf{T}_*, \tag{7.12}$$

and in terms of Φ by

$$\mathbf{E}_* = \mathbf{T}_*\mathbf{J}^{-\mathrm{t}}(\|\mathbf{J}^{-\mathrm{t}}\mathbf{n}\|^{-1}\mathbf{E} + \mathrm{D}^2_{\mathbf{n}_*}\Phi)\mathbf{J}^{-1}\mathbf{T}_*. \tag{7.13}$$

Proof: Let F be the signed distance function of $\mathbf{H}$, as above. First, differentiating $F_* = F \circ \Phi_*$ twice yields $\mathrm{D}^2F_* = \mathrm{D}\Phi_*^{\mathrm{t}}\,\mathrm{D}^2F\,\mathrm{D}\Phi_* + \mathrm{D}^2_{\mathrm{D}F}\Phi_*$. Since F is assumed to be the signed distance function, $\mathrm{D}^{\mathrm{t}}F = \mathbf{N}$ and $\mathrm{D}^2F = \mathbf{E}$. Further, $\mathrm{D}^{\mathrm{t}}F_* = \mathbf{J}_*^{\mathrm{t}}\mathrm{D}^{\mathrm{t}}\mathbf{F} = \mathbf{J}_*^{\mathrm{t}}\mathbf{N}$, and comparison with (7.10) proves (7.12). Second, differentiating $F = F_* \circ \Phi$ twice yields

$$\mathbf{E} = \mathrm{D}^2F = \mathrm{D}\Phi^{\mathrm{t}}\,\mathrm{D}^2F_*\,\mathrm{D}\Phi + \mathrm{D}^2_{\mathrm{D}F_*}\Phi = \mathbf{J}^{\mathrm{t}}\,\mathrm{D}^2F_*\,\mathbf{J} - \|\mathrm{D}F_*\|\,\mathrm{D}^2_{\mathbf{n}_*}\Phi.$$

Hence,

$$\|\mathbf{J}^{-\mathrm{t}}\mathbf{n}\|^{-1}\,\mathrm{D}^2F_* = \mathbf{J}^{-\mathrm{t}}\,(\|\mathbf{J}^{-\mathrm{t}}\mathbf{n}\|^{-1}\,\mathbf{E} + \mathrm{D}^2_{\mathbf{n}_*}\Phi)\mathbf{J}^{-1},$$

and, in view of (7.10), multiplication with T_* from the left and the right verifies (7.13). □

If $\Phi : \mathbf{x} \to \mathbf{A}\mathbf{x} + \mathbf{x}_0$ is an invertible affine map, then formulas (7.12) and (7.13) attain a particularly simple form,

$$\mathbf{E}_* = \|\mathbf{A}^{-\mathrm{t}}\mathbf{n}\|^{-1}\mathbf{T}_*\mathbf{A}^{-\mathrm{t}}\mathbf{E}\mathbf{A}^{-1}\mathbf{T}_*.$$

Unfortunately, the formulas derived above cannot be used for the ambient B-spline method because the obtained spline approximates a general function in $\mathbb{R}^d$ that has been extended constantly into normal direction which means that the obtained spline or the function Φ in the above context usually generates singular points and has small normal derivatives. Since the derived formulas in Theorem 7.5 use the inverse of the Jacobian, they are not the right tool for the surface examples in this thesis. Therefore our goal is to derive a numerically robust formula for the computation of the normal vector $\mathbf{n}_*$ and the curvature tensor $\mathbf{E}_*$ in the setting of the ambient B-spline method. The Jacobian of the given approximant Φ is again denoted by

$$\mathbf{J} := \mathrm{D}\Phi.$$

The tangent space on $\mathbf{H}_*$ is generated by the columns of $\mathbf{J}$ when restricted to the tangent space on $\mathbf{H}$. We therefore assume that the eigenvalues $\mu_1, \ldots, \mu_d$ of the matrix $\mathbf{J}$ fulfill the following: $\|\mu_1\| \geq \cdots \geq \|\mu_{d-1}\| \gg \|\mu_d\|$. The better the underlying approximation, the smaller $\|\mu_d\|$ and if we dealt with the exact function of our extension, we would get $\mu_d = 0$.

Choosing $d-1$ linearly independent tangent vectors $\mathbf{t}_1, \ldots, \mathbf{t}_{d-1}$ of $\mathbf{H}$, we know that $\mathbf{J}\mathbf{t}_1, \ldots, \mathbf{J}\mathbf{t}_{d-1}$ are linearly independent tangent vectors of $\mathbf{H}_*$. We choose $\mathbf{t}_1, \ldots, \mathbf{t}_{d-1}$ in such a way, that $(\mathbf{t}_1, \ldots, \mathbf{t}_{d-1}, \mathbf{n})$ are positively oriented, i.e. $\det([\mathbf{t}_1, \ldots, \mathbf{t}_{d-1}, \mathbf{n}]) > 0$, and compute the normal vector $\mathbf{n}_*$ as the cross product of the $d-1$ images $\mathbf{J}\mathbf{t}_1, \ldots, \mathbf{J}\mathbf{t}_{d-1}$ divided by its norm:

$$\mathbf{n}_* = \frac{\mathbf{J}\mathbf{t}_1 \times \cdots \times \mathbf{J}\mathbf{t}_{d-1}}{\|\mathbf{J}\mathbf{t}_1 \times \cdots \times \mathbf{J}\mathbf{t}_{d-1}\|}. \tag{7.14}$$

Theorem 7.6. *Let $\boldsymbol{H} \subset \mathbb{R}^d$ be a hypersurface with curvature tensor $\boldsymbol{E}$, and $\boldsymbol{H}_* := \Phi(\boldsymbol{H})$ with $\Phi \in C^2(\mathbb{R}^d)$ as introduced above. Then, with $\boldsymbol{T} := \mathrm{Id} - \boldsymbol{n}\boldsymbol{n}^{\mathrm{t}}$, the $d \times (d+1)$ auxiliary matrix $\boldsymbol{R} := \begin{bmatrix}\boldsymbol{J}\boldsymbol{T} & \boldsymbol{n}_*\end{bmatrix}$ and the notation introduced above, the curvature tensor of $\boldsymbol{H}_*$ is given in terms of Φ by*

$$\boldsymbol{E}_* = (\boldsymbol{R}\boldsymbol{R}^{\mathrm{t}})^{-1}\boldsymbol{J}\boldsymbol{T}\left(\|\boldsymbol{J}^{\mathrm{t}}\boldsymbol{n}_*\|\boldsymbol{E} + \boldsymbol{T}\,\mathrm{D}^2_{\boldsymbol{n}_*}\Phi\,\boldsymbol{T}\right)\boldsymbol{T}\boldsymbol{J}^{\mathrm{t}}(\boldsymbol{R}\boldsymbol{R}^{\mathrm{t}})^{-1}. \tag{7.15}$$

Proof: Let F_* be the signed distance function of $\mathbf{H}_*$, i.e. $F_{*|\mathbf{H}_*} = 0$, $\mathrm{D}F_{*|\mathbf{H}_*} = -\mathbf{n}_*$ and $\|\mathrm{D}F_*\| = 1$. Then $F := F_* \circ \Phi$ is an implicit representation of $\mathbf{H}$. Differentiating $F := F_* \circ \Phi$ twice yields

$$\mathrm{D}^2 F = \mathbf{J}^{\mathrm{t}}\,\mathrm{D}^2 F_*\,\mathbf{J} + \mathrm{D}^2_{-\mathbf{n}_*}\Phi.$$

Since F_* is assumed to be the signed distance function, we know $\mathrm{D}^2 F_* = \mathbf{E}_*$ and obtain

$$\mathrm{D}^2 F = \mathbf{J}^{\mathrm{t}} \mathbf{E}_* \mathbf{J} - \mathrm{D}^2_{\mathbf{n}_*} \Phi. \tag{7.16}$$

We know that $\mathbf{E} = \frac{1}{\|\mathrm{D}F\|} \mathbf{T}\,\mathrm{D}^2 F \mathbf{T}$ for $\mathbf{T} = \mathrm{Id} - \mathbf{n}\mathbf{n}^{\mathrm{t}}$ and $\|\mathrm{D}F\| \neq 0$ on $\mathbf{H}$. But in case we have a non regular implicit function with $\|\mathrm{D}F\| = 0$, we easily show that $\|\mathrm{D}F\|\mathbf{E} = \mathbf{T}\,\mathrm{D}^2 F \mathbf{T}$ is still true. Let $\mathbf{f}$ be a regular parametrization of $\mathbf{H}$. Differentiating $F \circ \mathbf{f} = 0$ in arbitrary directions i and j reveals

$$\mathrm{d}\mathbf{f}_j^{\mathrm{t}} (\mathrm{D}^2 F \circ \mathbf{f})\, \mathrm{d}\mathbf{f}_i + (\mathrm{D}F \circ \mathbf{f})\, \mathrm{d}^2 \mathbf{f}_{i,j} = 0.$$

The second summand is zero by assumption. Thus, we conclude that $(\mathrm{D}^2 F \circ \mathbf{f})$ is either zero or has only one eigenvalue not equal to zero with corresponding eigenvector $\mathbf{n}$. But this is equivalent to $\mathbf{T}\,\mathrm{D}^2 F \mathbf{T} = 0$.

By the chain rule, we know $\mathrm{D}F = \mathrm{D}F_* \mathbf{J}$ and therefore $\|\mathrm{D}F\| = \|\mathbf{J}^{\mathrm{t}} \mathbf{n}_*\|$. We therefore obtain with equation (7.16)

$$\begin{aligned} \|\mathbf{J}^{\mathrm{t}}\mathbf{n}_*\|\mathbf{E} &= \mathbf{T}\left(\mathbf{J}^{\mathrm{t}} \mathbf{E}_* \mathbf{J} - \mathrm{D}^2_{\mathbf{n}_*} \Phi\right) \mathbf{T}. \\ \Leftrightarrow \qquad \|\mathbf{J}^{\mathrm{t}}\mathbf{n}_*\|\mathbf{E} + \mathbf{T}\,\mathrm{D}^2_{\mathbf{n}_*} \Phi \mathbf{T} &= \mathbf{T}\mathbf{J}^{\mathrm{t}} \mathbf{E}_* \mathbf{J}\mathbf{T}. \end{aligned}$$

With $\mathbf{J}^{\mathrm{t}}\mathbf{n}_* = \alpha\mathbf{n}$, $\alpha \in \mathbb{R}$, we conclude that $\mathbf{T}\mathbf{J}^{\mathrm{t}}\mathbf{n}_* = \mathbf{T}\alpha\mathbf{n} = \mathbf{0}$ holds, independent of the value α. But this reveals that $\mathbf{n}_*$ is eigenvector to the eigenvalue zero of the matrix $\mathbf{T}\mathbf{J}^{\mathrm{t}}$. Knowing that $\operatorname{rank}(\mathbf{T}\mathbf{J}^{\mathrm{t}}) = d - 1$ holds, we extend the above equation as follows:

$$\begin{pmatrix} \|\mathbf{J}^{\mathrm{t}}\mathbf{n}_*\|\mathbf{E} + \mathbf{T}\,\mathrm{D}^2_{\mathbf{n}_*} \Phi \mathbf{T} & \mathbf{0} \\ \mathbf{0} & 0 \end{pmatrix} = \begin{pmatrix} \mathbf{T}\mathbf{J}^{\mathrm{t}} \\ \mathbf{n}_*^{\mathrm{t}} \end{pmatrix} \mathbf{E}_* \underbrace{\begin{pmatrix} \mathbf{J}\mathbf{T} & \mathbf{n}_* \end{pmatrix}}_{=:\mathbf{R}},$$

where we use $\mathbf{E}_*\mathbf{n}_* = \mathbf{0}$ and refer to $\mathbf{0}$ as the zero column or row, which is needed. Defining the auxiliary matrix $\mathbf{R} := \begin{bmatrix} \mathbf{J}\mathbf{T} & \mathbf{n}_* \end{bmatrix}$, we multiply the last equation from the left by $\mathbf{R}$ and from the right by $\mathbf{R}^{\mathrm{t}}$ and obtain

$$\mathbf{J}\mathbf{T}\left(\|\mathbf{J}^{\mathrm{t}}\mathbf{n}_*\|\mathbf{E} + \mathbf{T}\,\mathrm{D}^2_{\mathbf{n}_*} \Phi \mathbf{T}\right) \mathbf{T}\mathbf{J}^{\mathrm{t}} = \mathbf{R}\mathbf{R}^{\mathrm{t}} \mathbf{E}_* \mathbf{R}\mathbf{R}^{\mathrm{t}}.$$

The $d \times d$ matrix $\mathbf{R}\mathbf{R}^{\mathrm{t}}$ is invertible and we finally verify formula (7.15)

$$\mathbf{E}_* = (\mathbf{R}\mathbf{R}^{\mathrm{t}})^{-1} \mathbf{J}\mathbf{T}\left(\|\mathbf{J}^{\mathrm{t}}\mathbf{n}_*\|\mathbf{E} + \mathbf{T}\,\mathrm{D}^2_{\mathbf{n}_*} \Phi \mathbf{T}\right) \mathbf{T}\mathbf{J}^{\mathrm{t}} (\mathbf{R}\mathbf{R}^{\mathrm{t}})^{-1}.$$

□

Equation (7.15) neither contains an inverse of $\mathbf{J}$ nor is divided by a number that may become small during an enhanced approximation process and is therefore numerically robust. With the help of the last derived formula, we investigate in the upcoming section the differential geometric quantities such as normal vectors and principal curvatures not only for the examples of the Stanford Bunny and the cow model but also for a rather academic problem for which we have access to the true underlying curvature information.

7.4 Examples

With the help of the theoretical considerations of the previous section, we want to numerically determine differential geometric quantities of the approximated target surfaces as well as investigate how well these can be computed. For this, we first look at the already examined examples for which we have no reference data. As already mentioned in Subsection 7.3.1, a common way to obtain normal vectors and curvature information on faceted surfaces could be the usage of a discrete scheme. Nevertheless, with the help of the ambient B-spline method we may derive these quantities analytically. We additionally determine differential geometric quantities on an algebraic surface in order to numerically verify expected rates of convergence.

7.4.1 Stanford Bunny and cow model

In this subsection we demonstrate the results of Equations (7.14) and (7.15) when applied to the two example surfaces we already investigated. Figure 7.1 illustrates normals at some vertices of the Stanford Bunny. The spline approximant modeling the two surface examples depends to some extent on the underlying triangulation because the value assignment of the data sites' footpoints uses the latter, cp. Section 4.2. Nevertheless, in contrast to the discrete setting in which adverse triangulations may result in bad estimations of the normal vectors, the most famous example is probably the Schwarz lantern introduced in [Sch90], the generated normals by (7.14) indeed correspond to an existing surface which approximates the given data. The normal vectors of the approximation of the cow model can be evaluated equivalently.

Figures 7.2 and 7.3 show the images of spline approximants colored by the corresponding Gaussian curvature obtained with the help of (7.15). We recognize that neither of the two example surfaces comprises a nice curvature distribution due to existing noise as well as the surface details such as the structure of the bunny's fur. However, since discrete estimation schemes for curvature information are generally even more sensitive to noise within the data or adverse triangulations than for normal vectors, the ambient B-spline method easily provides access to an approximation of curvature information that indeed belongs to an analytic surface which can sufficiently approximate the original data.

Figure 7.1: The triangulated image of the spline approximant of order $n = 4$ and grid size $h = 0.05$. The figures show randomly chosen and red marked normal vectors computed with the help of (7.14).

7.4.2 Rounded cube

We investigate on a simple algebraic surface in $\mathbb{R}^3$ for which we can easily establish a one-to-one mapping to the unit sphere: a cube with rounded edges and corners. Such a surface is the unit ball of the p-norm and is implicitly given by the following equation:

$$C := \{x \in \mathbb{R}^3 \mid |x_1|^p + |x_2|^p + |x_3|^p = 1\}, \quad p > 2. \tag{7.17}$$

A point $y \in \mathbb{R}^3$ on the unit sphere can be uniquely mapped onto such a surface by finding the scaling factor t such that

$$y + t \cdot y \in C, \quad t \geq 0.$$

If we substitute the latter into the implicit equation of the regarded cube we obtain

$$t = \left(\frac{1}{|y_1|^p + |y_2|^p + |y_3|^p} \right)^{\frac{1}{p}} - 1.$$

The greater the parameter p the less the edges and corners are rounded, i.e. in the limit for $p \to \infty$ one obtains a cube with sharp corners and edges. Since we want to approximate normal vectors and curvature information, we choose the rather moderate number $p = 8$ for which we can also neglect the absolute values in formula (7.17). In Subsection 7.3.2 we already introduced how to obtain normal vectors and curvature information for implicit surfaces, therefore these quantities are easily computed for a comparison.

We determine approximations via the ambient B-spline method by bisecting the grid sizes ranging from 0.32 to 0.02 in order to verify the estimated rate of convergence for the approximation error of the data, the normal vectors as well as the curvature information. The used B-spline order equals $n = 4$. Although the shape of the underlying example is rather simple, since it is convex everywhere, i.e. there are no hyperbolic points on the surface, its modeling is a challenging task and the used B-splines indeed produce hyperbolic points as is illustrated in Figure 7.4. Nevertheless, this example does confirm the predicted convergence rates as can be seen in Figures 7.5 - 7.7.

Figure 7.5 shows the maximal and the root mean squares deviations from the image to the original set C. Accordingly, Figure 7.6 represents the distances between the normal vectors of the spline approximant and the true normal vectors. The deviations in the curvature information in Figure 7.7 are measured in the Frobenius norm of the differences of the true curvature tensors and the curvature tensors computed by the spline approximant. Figures 7.5 and 7.7 deal with the relative errors, that means the absolute deviations are scaled by a factor which relates the error to the absolute values of the underlying data.

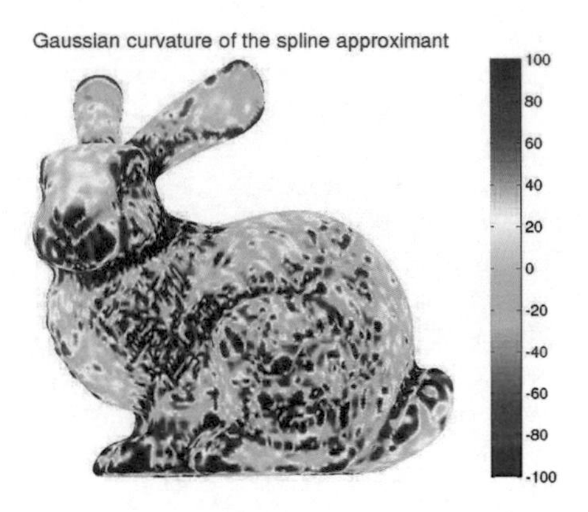

Figure 7.2: The image of the spline approximant of order $n = 4$ and grid size $h = 0.05$ colored by its Gaussian curvature obtained by Equation (7.15).

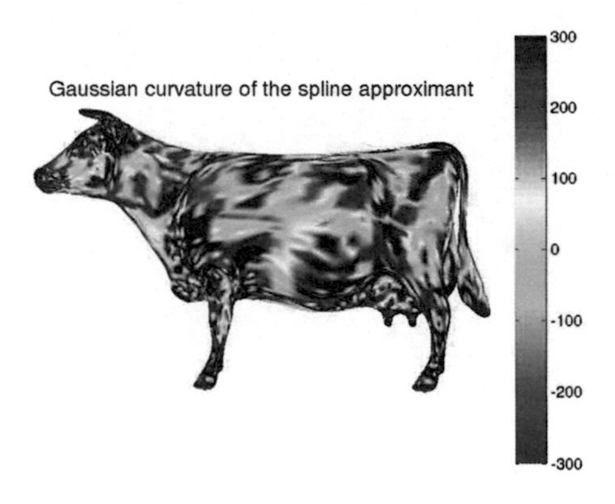

Figure 7.3: The image of the spline approximant of order $n = 4$ and grid size $h = 0.02$ colored by its Gaussian curvature obtained by Equation (7.15).

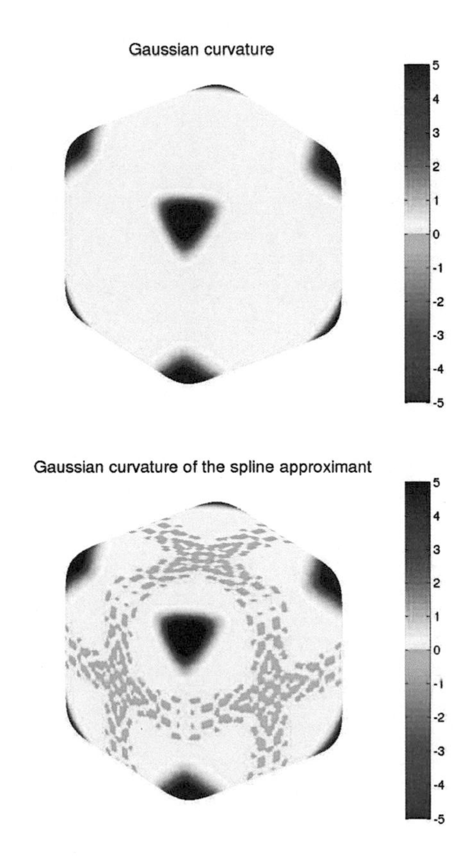

Figure 7.4: Comparison of true underlying and approximated Gaussian curvature of the example for grid size $h = 0.08$ and order $n = 4$. Whereas the original surface is convex, the discontinuous color gradient of the colorbar reveals hyperbolic points (cyan) of small negative Gaussian curvature of the spline approximant.

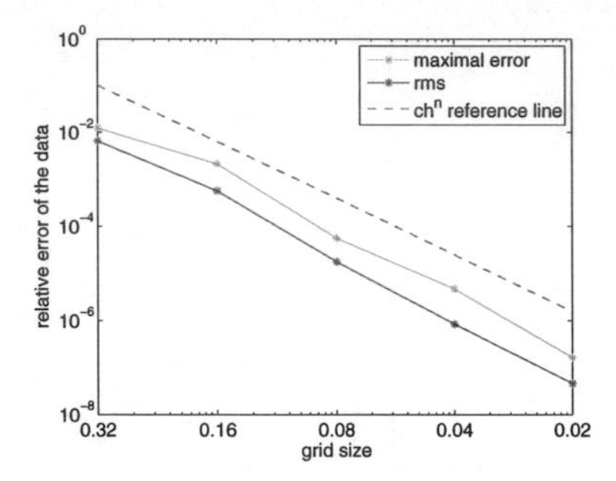

Figure 7.5: Experimental rate of convergence for the approximated data of C.

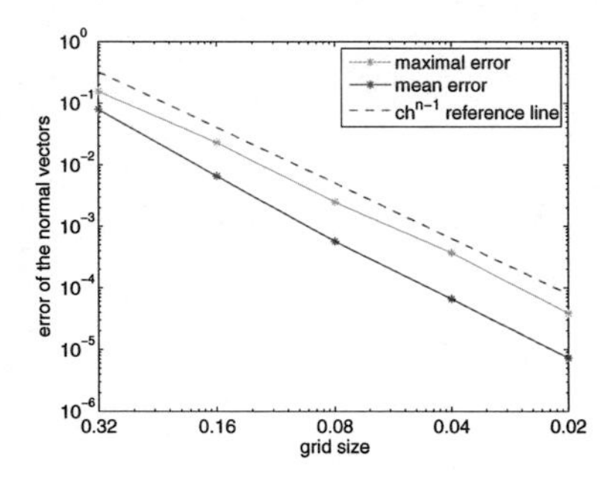

Figure 7.6: Experimental rate of convergence for the approximated normal vectors of C.

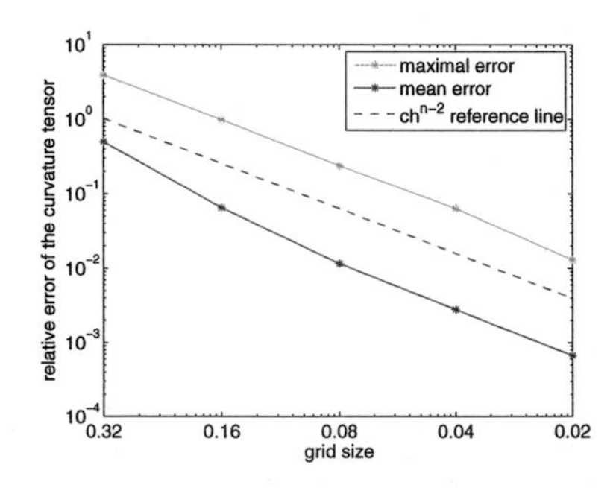

Figure 7.7: Experimental rate of convergence for the approximated curvature information of C.

8 Further Issues

We briefly introduce two topics which have been part of the research work but did not provide the desired benefits we looked for. The first concerns the idea of parameter correction which was introduced in [HL93]. The hope was to apply the latter to examples like the Stanford Bunny or the cow model, i.e. for polygonal surfaces which are mapped to a reference surface. As usual, a reasonable choice of the parameter positions is not canonical. Therefore the main idea is to iteratively generate error vectors that are perpendicular to the image of the spline s, although there is no guarantee or any theoretical result about convergence. The second topic with which we also tried to enhance the quality of the surface examples was the implementation of a so-called bootstrapping strategy. We understand the latter as a self-improving process or in our case an algorithm that uses old output data as new input data.

8.1 Parameter correction

We already mentioned that finding a spherical parametrization for genus zero objects such as the Stanford Bunny or the cow model is generally a difficult task, especially guaranteeing the bijectivity as well as the consistency of a polygonal mesh, cp. Chapter 2. The spherical parametrizations generated by the authors of [PH03] we use in this thesis are produced with much effort and we already generated nice results by the ambient B-spline method. Nevertheless, just because a model is bijectively mapped to a reference manifold does not indicate that the chosen parameters, in our examples the vertices on the sphere, are well placed. As explained in [HL93], the main idea of a so-called parameter correction is derived from the ansatz to minimize the error vector at a point $x \in \mathcal{M}$

$$\min_{\Delta x \in \mathbb{R}^d} \|g_x^{\text{fix}} - s(x + \Delta x)\|^2,$$

where g_x^{fix} denotes a fixed function value with initial preimage x and Δx is the parameter correction term to be determined. Analogous to the procedure of handling curves, we may approximate the entries of the vector Δx by separating the coordinate directions. Thus, differentiating equation $(g_x^{\text{fix}} - s(\ldots x_{i-1}, x_i + \Delta x_i, x_{i+1}, \ldots))^2$ into direction x_i leads to conditions

$$(g_x^{\text{fix}} - s(\ldots x_{i-1}, x_i + \Delta x_i, x_{i+1}, \ldots))^t \cdot s_{x_i}(\ldots x_{i-1}, x_i + \Delta x_i, x_{i+1}, \ldots) = 0, \qquad (8.1)$$

where s_{x_i}, $i = 1,\dots,d$ denotes the derivative of the spline s into coordinate direction x_i. With the help of Taylor expansions, we obtain the following two expressions:

$$s(\dots x_{i-1}, x_i + \Delta x_i, x_{i+1}, \dots) \approx s(x) + s_{x_i}(x) \cdot \Delta x_i$$

and

$$s_{x_i}(\dots x_{i-1}, x_i + \Delta x_i, x_{i+1}, \dots) \approx s_{x_i}(x) + s_{x_i,x_i}(x) \cdot \Delta x_i$$

respectively, where $s_{x_i,x_i}(x)$ indicates that s is differentiated twice into direction x_i. Substituting the two latter expansions into (8.1) and neglecting all terms that are at least quadratic, we obtain analogously to [HL93] a correction formula:

$$\Delta x_i = \frac{(g_x^{\mathrm{fix}} - s(x))^{\mathrm{t}} \cdot s_{x_i}(x)}{s_{x_i}(x)^{\mathrm{t}} s_{x_i}(x) - (g_x^{\mathrm{fix}} - s(x))^{\mathrm{t}} s_{x_i,x_i}(x)} \qquad i = 1,\dots,d. \tag{8.2}$$

For classic parametrized curves or surfaces, one updates a parameter x by setting $x^* = x + \varepsilon\Delta x$, ε reducing the update direction if necessary. The approach only works, if the quadratic terms are neglegtable. This requirement should be numerically checked before (8.2) is applied. Nevertheless, an update in our case turns out to be more complicated since we not only have to guarantee the embedding of the polygonal mesh, i.e. we do not want to create any foldovers, but also need to assure that an updated vertex x^* still lies on the reference manifold, which is generally wrong for $\Delta x \in \mathbb{R}^d$. We therefore experimented with the following update formula

$$x^* = \mathrm{clp}_{\mathcal{M}}(x + \varepsilon\Delta x), \ \varepsilon > 0,$$

where ε guarantees the embedding of the triangulation and the closest point projection to be well-defined. The step size may be determined heuristically via bisection. Another option could be an update formula which projects the point $x + \varepsilon\Delta x$ onto the reference manifold along the error vector $g_x^{\mathrm{fix}} - s(x + \varepsilon\Delta x)$. The latter idea does not change the angle between the error vector and the surface.

The experiments concerning parameter correction with the Stanford Bunny and cow model hardly produced any changes in the error behavior. We assume that this is due to two reasons: first, the spherical parametrizations we started with are already very good since they are generated by minimizing a stretch metric, cp. Chapter 2. Second, both models consist of very fine meshes that leave little leeway for parameter changes without creating triangle foldovers. Consequently, in case one starts with a coarse and more randomly generated embedding of the vertices on a reference manifold, some steps of a parameter correction method as described above may decrease the approximation error without the need of using a smaller grid size but we could not experimentally reassess this assumption.

8.2 Bootstrapping

The concept of bootstrapping approaches arises in many different fields[1] and is often used as a metaphor for a self-sustaining or -generating process. The term presumably bases on the famous tall tale of the Baron Münchhausen who allegedly pulled himself out of a swamp by his own hair. The idiomatic expression "to pull oneself up by one's bootstraps" arose possibly of a misattribution or because of a propagated variant of the original story.

A spline approximant as a result of the ambient B-spline method that maps vertices of a reference manifold to a target manifold like the Stanford Bunny or the cow model inevitable contains high distortion concerning the area and in case we start with a less carefully generated spherical parametrization also in the polygon's angles. Since the approximation error depends on all derivatives of the function to be approximated, cp. Theorem 3.3 of Section 3.2, there is a reasonable suspicion that high derivative norms caused by distortion may have a negative impact on the approximation error. This consideration raises the question whether a composition of some spline approximations, each containing less distortion than the primary mapping, may lead to a decrease in the approximation error. The bootstrapping approach which we investigated is illustrated in Figure 8.1 with the help of the Stanford Bunny example: we replace the direct approximation s from the sphere to a nearly perfect bunny model by the composition of k spline functions. First, the spline s_0 approximates a highly smoothed model which lacks any features but resembles the target surface better than the original reference manifold. Second, the spline s_1 uses the image of the approximant s_0 as new reference manifold and applies the ambient B-spline method to a less smoothed model in order to obtain more details of the underlying geometric model. Theoretically, one could repeat this step a couple of times until the last iteration finally generates a spline s_k which approximates the original data with the image of the spline s_{k-1} as reference manifold. In such an iterative process, the last spline approximant s_k restricted to its corresponding reference manifold should almost be the identity because its reference manifold and the target manifold differ only by little. The identity in $\mathbb{R}^d$ can generally be reproduced by splines, we therefore hoped for an overall improvement of the approximation. Unfortunately, the identity on a manifold extended constantly into its ambient space is a nontrivial function and therefore not easily reproduced by splines. However, one sticking point of the suggested bootstrapping idea is that even highly smoothed data may locally create high principal curvatures which may force the grid size in an upcoming iteration step to be disproportional small compared to the underlying data density and its information content. Secondly, as already pointed out in Section 3.1, the existence of two connected components that may be close in the Euclidean space such as the two ears of the Stanford Bunny also force the ambient B-spline method to decrease the grid size around these areas. Consequently, the immense increase

[1] i.e. computer science, biology, physics, statistics and others

of the degrees of freedom results into a huge increase of the computational costs for the local least squares fits as well as for the determination of the closest points of the data sites. Finally, we deduce that a direct approximation on an initial, geometrically simple reference manifold with appropriate hierarchical grid sizes delivers good approximation results without causing out of scale computational costs.

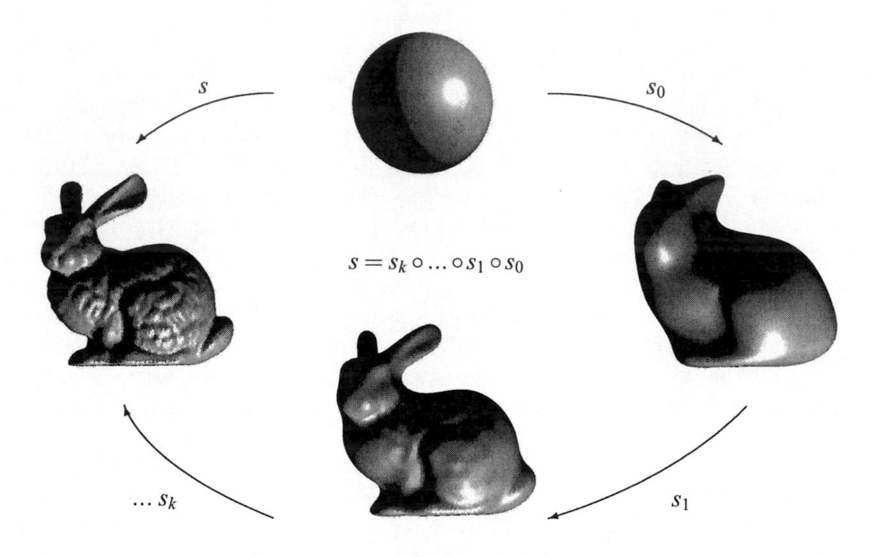

Figure 8.1: Illustration of the bootstrapping idea.

9 Conclusion and Outlook

This thesis treats a new approach for the approximation of functions defined on manifolds: the ambient B-spline method. This method is independent of the manifold's genus and representation and easily generates C^k-approximants. The key ingredient is the constant extension of the function into normal directions of the manifold which detaches us from defining complicated function spaces on the manifold itself. The proceeding then is an approximation in the manifold's ambient space, i.e. on subdomains in $\mathbb{R}^d$.

When a surface is bijectively mapped to a topologically equivalent surface, one may interpret each coordinate of the surface as function over its reference manifold. This means the method turns out to be a tool for modeling surfaces as soon as a bijective mapping to a reference manifold is at hand. A careful definition of the approximation domain enables us to use simple tensor product B-splines without generating singularities or facing stability problems.

First, we introduced the ambient B-spline method by focusing on theoretical aspects of the extension process. Additionally, using spline spaces for the approximation step provides full approximation power, although the approximation error of a function g on a manifold $\mathcal{M}$ depends on the Sobolev norm $\|g\|_{W_p^n(\mathcal{M})}$ in contrast to classic error estimates on domains $\Omega \subset \mathbb{R}^d$ which usually depend on the Sobolev semi-norm $|g|_{W_p^n(\Omega)}$.

We described in detail the procedure for an algorithmic implementation with the help of an illustrative example in $\mathbb{R}^2$. The three main examples of the thesis, the geoid, the Stanford Bunny and the cow model, use the unit sphere as reference manifold and reveal good approximation results but also demonstrate the necessity for a multiscale procedure.

We established two alternatives for adaptive approximation techniques: we chose hierarchical overlays for the very irregular distributions of the geoid undulations and hierarchical B-splines for the two surface models. Both techniques are presented and discussed. Due to the restrictive refinement rule of our second alternative, we established a stable, hierarchical basis in contrast to the setting treated in the thesis [Kra98].

The recalculation of the main examples by hierarchical approaches revealed promising results: we obtained a globally consistent approximation of geoid heights which could be easily updated due to the local support of B-spline functions. However, the price to pay was the general increase in the number of coefficients due to the extra dimension. Furthermore, the hierarchical approach was able to detect complicated regions of the Stanford Bunny and the cow model and concentrated on high resolutions in areas of detailed features such as the bunny's ears or the cow's legs.

Concerning the modeling of surfaces, in this thesis we recapitulated some differential geometric quantities in different settings and then introduced a numerically robust way to calculate curvature information of the target surface as a function of curvature information of the reference manifold. Subsequently, we investigated the derived formulas with the help of two surface examples. Together with the ambient B-spline approach, the formulas provide a possibility to analytically estimate normal vectors and curvature information of a former discrete model. Secondly, with the help of an algebraic surface for which we could compute the exact higher order quantities, we confirmed the approximation error behavior of points, normal vectors and curvature information when the ambient B-spline method is applied. It turned out that the reference manifolds should have a limited and uniform curvature distribution because high principal curvatures may force used grid sizes to become uneconomically small.

This thesis only analyzed examples calculated on the sphere or the circle, respectively. An approximation of a function on a general manifold is just as straightforward if we do not treat reference manifolds that have a too uneven curvature distribution or two connected components which are close in the Euclidean sense, cf. Chapter 3. As long as the reference manifold is at least C^2 the method theoretically works fine but may cause undesirably many degrees of freedom as we observed investigating the bootstrapping approach.

The ambient method provides a nice possibility of modeling surfaces but clearly has an unpleasant bottleneck: the generation of a bijective mapping of an initial mesh to its topologically equivalent reference manifold is numerically expensive and for surfaces of genus greater than zero not satisfactorily analyzed and solved, especially concerning computation time. The ambient B-spline method works best if surface features get disproportionally more area on the reference manifold than they actually cover in the target manifold. The spherical parametrization provided by [PH03] is generated with much care concerning the angles of the triangulation. We assume that such a careful treatment is not necessary for the ambient B-spline method to produce satisfactory results as long as feature vertices do not contract on a too small surface area.

A future project could work on finding a reasonable heuristic approach to bijectively mapping meshes with arbitrary genus to corresponding reference manifolds. Therewith arises the question which reference manifolds one should choose for genus greater than one. For instance, there is no simple explicit parametrization for a surface with two holes such as a torus for the genus one example. Nevertheless, the used reference manifolds only need to satisfy a certain smoothness condition as well as to have manageable principle curvatures to prevent the grid size to become too small. As already mentioned in Chapter 2, subdivision schemes that can be analytically analyzed usually produce singularities at extraordinary vertices or flat points, cf. [PU98] and references therein. Though the latter may lead to bounded curvatures, one often observes concentric wave patterns which are unwanted when seeking high quality surfaces, cf. [KPR04]. However, as long as these wave patterns are bounded, they would not tremendously interfere with the ambient B-spline approach. Consequently, a future work should identify appropriate reference manifolds and provide explicit parametrizations.

Bibliography

[Ada75] Robert A. Adams. *Sobolev Spaces*. Academic Press, 1975.

[AH12] Kendall Atkinson and Weimin Han. *Spherical Harmonics and Approximations on the Unit Sphere: An Introduction*. Springer, 2012.

[Ahn04] Sung Joon Ahn. *Least Squares Orthogonal Distance Fitting of Curves and Surfaces in Space*. Springer, 2004.

[ANS96] Peter Alfeld, Marian Neamtu, and Larry L. Schumaker. Bernstein-Bézier Polynomials on Spheres and Sphere-Like Surfaces. *Comput. Aided Geom. Design*, 13:333–349, 1996.

[Bar09] Franz Barthelmes. Definition of Functionals of the Geopotential and Their Calculation from Spherical Harmonic Models. Technical Report STR09/02, German Research Center For Geosciences, Potsdam, Germany, February 2009. Theory and formulas used by the calculation service of the International Centre for Global Earth Models (ICGEM), http://icgem.gfz-potsdam.de.

[BH70] James H. Bramble and Stephen R. Hilbert. Estimation of Linear Functionals on Sobolev Spaces with Application to Fourier Transforms and Spline Interpolation. *SIAM Journal on Numerical Analysis*, 7(1):112–124, 1970.

[BLS05] Victoria Baramidze, Ming-Jun Lai, and C. K. Shum. Spherical Splines for Data Interpolation and Fitting. *SIAM J. Scientific Computing*, 28:241–259, 2005.

[BMZB05] Ioana Boier-Martin, Denis Zorin, and Fausto Bernardini. A Survey of Subdivision-Based Tools for Surface Modeling. DIMACS Series in Discrete Mathematics and Theoretical Computer Science, 2005.

[BOP92] Robert E. Barnhill, Karsten Opitz, and Helmut Pottmann. Fat Surfaces: a Trivariate Approach to Triangle-Based Interpolation on Surfaces. *Computer Aided Geometric Design*, 9(5):365 – 378, 1992.

[BPJ08] Janine Bennett, Valerio Pascucci, and Kenneth Joy. A Genus Oblivious Approach to Cross Parameterization. *Comput. Aided Geom. Des.*, 25(8):592–606, November 2008.

[BPK98] Alexander G. Belyaev, Alexander A. Pasko, and Tosiyasu L. Kunii. Ridges and Ravines on Implicit Surfaces. In *Proceedings of the Computer Graphics International 1998*, CGI '98, pages 530–, Washington, DC, USA, 1998. IEEE Computer Society.

[Bro65] Charles G. Broyden. A Class of Methods for Solving Nonlinear Simultaneous Equations. *Mathematics of Computation*, 19:577–593, 1965.

[BW97] Jules Bloomenthal and Brian Wyvill, editors. *Introduction to Implicit Surfaces*. Morgan Kaufmann Publishers Inc., San Francisco, CA, USA, 1997.

[Cas10] Thomas J. Cashman. *NURBS-Compatible Subdivision Surfaces*. PhD thesis, University of Cambridge, Computer Laboratory, Queens College, 2010.

[CCG+06] Cláudio G. S. Cardoso, Maria Cristina Cunha, Anamaria Gomide, Denis J. Schiozer, and Jorge Stolfi. Finite Elements on Dyadic Grids with Applications. *Math. Comput. Simul.*, 73(1):87–104, November 2006.

[CLCQ12] Juan Cao, Xin Li, Zhonggui Chen, and Hong Qin. Spherical DCB-Spline Surfaces with Hierarchical and Adaptive Knot Insertion. *IEEE Transactions on Visualization and Computer Graphics*, 18(8):1290–1303, August 2012.

[CSM03] David Cohen-Steiner and Jean-Marie Morvan. Restricted Delaunay Triangulations and Normal Cycle. In *Symposium on Computational Geometry*, pages 312–321, 2003.

[dB73] Carl de Boor. The Quasi-Interpolant as a Tool in Elementary Polynomial Spline Theory, 1973.

[dB78] Carl de Boor. *A Practical Guide to Splines*. Springer, 1978.

[DBD+13] Bailin Deng, Sofien Bouaziz, Mario Deuss, Juyong Zhang, Yuliy Schwartzburg, and Mark Pauly. Exploring Local Modifications for Constrained Meshes. *Computer Graphics Forum (Proceedings of Eurographics 2013)*, 32(2), 2013. to appear.

[dBF73] Carl de Boor and George J. Fix. Splineapproximation by Quasiinterpolants. *Journal of Approximation Theory*, 8(1):19–45, 1973.

[DKT98] Tony DeRose, Michael Kass, and Tien Truong. Subdivision Surfaces in Character Animation. In *Proceedings of the 25th annual conference on Computer graphics and interactive techniques*, SIGGRAPH '98, pages 85–94, New York, NY, USA, 1998. ACM.

[DLP13] Tor Dokken, Tom Lyche, and Kjell Fredrik Pettersen. Polynomial Splines over Locally Refined Box-partitions. *Computer Aided Geometric Design*, 30(3):331–356, 2013.

[DPR13] Oleg Davydov, Jennifer Prasiswa, and Ulrich Reif. Two-Stage Approximation Methods with Extended B-Splines. *Mathematics of Computation*, 2013.

[DS07] Oleg Davydov and Larry L. Schumaker. Scattered Data Fitting on Surfaces Using Projected Powell-Sabin Splines. In *Proceedings of the 12th IMA international conference on Mathematics of surfaces XII*, pages 138–153, Berlin, Heidelberg, 2007. Springer-Verlag.

[DVJK08] Giovanni Della Vecchia, Bert Jüttler, and Myung-Soo Kim. A Construction of Rational Manifold Surfaces of Arbitrary Topology and Smoothness from Triangular Meshes. *Comput. Aided Geom. Des.*, 25(9):801–815, December 2008.

[EW05] Jeff Erickson and Kim Whittlesey. Greedy Optimal Homotopy and Homology Generators. In *Proceedings of the sixteenth annual ACM-SIAM symposium on Discrete algorithms*, SODA '05, pages 1038–1046, Philadelphia, PA, USA, 2005. Society for Industrial and Applied Mathematics.

[FB88] David R. Forsey and Richard H. Bartels. Hierarchical B-spline refinement. *SIGGRAPH Comput. Graph.*, 22(4):205–212, June 1988.

[FB95] David R. Forsey and Richard H. Bartels. Surface Fitting with Hierarchical Splines. *ACM Transactions on Graphics*, 14:134–161, 1995.

[Foo84] Robert L. Foote. Regularity of the Distance Function. *Proceedings of The American Mathematical Society*, 92:153–153, 1984.

[GGS03] Craig Gotsman, Xianfeng Gu, and Alla Sheffer. Fundamentals of Spherical Parameterization for 3D Meshes. *ACM Trans. Graph.*, 22(3):358–363, July 2003.

[GH03] Cindy Grimm and John Hughes. Parameterizing N-holed Tori. In *Mathematics of Surfaces X*, pages 14–29, September 2003.

[GHQ06] Xianfeng Gu, Ying He, and Hong Qin. Manifold splines. *Graph. Models*, 68(3):237–254, May 2006.

[GJPH09] Cindy Grimm, Tao Ju, Ly Phan, and John Hughes. Adaptive Smooth Surface Fitting with Manifolds. *Vis. Comput.*, 25(5-7):589–597, April 2009.

[Gol04] Jack Goldfeather. A Novel Cubic-order Algorithm for Approximating Principal Direction Vectors. *ACM Transactions on Graphics*, 23:45–63, 2004.

[Gol05] Ron Goldman. Curvature Formulas for Implicit Curves and Surfaces. *Comput. Aided Geom. Des.*, 22:632–658, October 2005.

[Gre03] Robin Green. Spherical Harmonic Lighting: The Gritty Details. 2003. Game Developers Conference.

[GVJ+09] Abel Gomes, Irina Voiculescu, Joaquim Jorge, Brian Wyvill, and Callum Galbraith, editors. *Implicit Curves and Surfaces: Mathematics, Data Structures and Algorithms*. Springer, 2009.

[GWC+04] Xianfeng Gu, Yalin Wang, Tony F. Chan, Paul M. Thompson, and Shing tung Yau. Genus Zero Surface Conformal Mapping and its Application to Brain Surface Mapping. *IEEE Transactions on Medical Imaging*, 23:949–958, 2004.

[GZ06] Cindy Grimm and Denis Zorin. Surface Modeling and Parameterization with Manifolds: Siggraph 2006 Course Notes Author presenation videos are available from the citation page . In *ACM SIGGRAPH 2006 Courses*, SIGGRAPH '06, pages 1–81, New York, NY, USA, 2006. ACM.

[HG00] Andreas Hubeli and Markus Gross. A Survey of Surface Representations for Geometric Modeling. Technical report, Institute of Scientific Computing, 2000. Technical Report 335, ETH Zürich.

[HL93] Josef Hoschek and Dieter Lasser. *Fundamentals of Computer Aided Geometric Design*. A. K. Peters, Ltd., Natick, MA, USA, 1993.

[Höl03] Klaus Höllig. *Finite Element Method with B-Splines*. Society for Industrial and Applied Mathematics, 2003.

[Hor97] Kai Hormann. Glatte Approximation mit hierarchischen Splineflächen. Master's thesis, Department of Mathematics, University of Erlangen, 1997.

[HP11] Klaus Hildebrandt and Konrad Polthier. Generalized Shape Operator on Polyhedral Meshes. *Computer Aided Geometric Design*, 2011.

[HPS08] Kai Hormann, Konrad Polthier, and Alia Sheffer. Mesh Parameterization: Theory and Practice. In *ACM SIGGRAPH ASIA 2008 courses*, SIGGRAPH Asia '08, pages 12:1–12:87, New York, NY, USA, 2008. ACM.

[HRW01] Klaus Höllig, Ulrich Reif, and Joachim Wipper. Weighted Extended B-Spline Approximation of Dirichlet Problems. *SIAM J. Numer. Anal.*, 39(2):442–462, February 2001.

[HW05] Shi-min Hu and Johannes Wallner. A Second Order Algorithm for Orthogonal Projection onto Curves and Surfaces. *Comp. Aided Geom. Des*, 22:251–260, 2005.

[HWW+06] Ying He, Kexiang Wang, Hongyu Wang, Xianfeng Gu, and Hong Qin. Manifold T-spline. In *In Proceedings of Geometric Modeling and Processing*, pages 409–422, 2006.

[Kan07] Christian Kanzow. Nichtlineare Gleichungssysteme. Julius Maximilians Universität Würzburg, Vorlesungsskript, 2007. Available online at `http://www.mathematik.uni-wuerzburg.de/~kanzow/ne/NE_07.pdf`; visited on August 8th 2013.

[Kno06] Aaron Knoll. A Short Survey of Octree Volume Rendering Techniques. GI Lecture Notes in Informatics. Proceedings of 1st IRTG Workshop, June 14-16 2006, Dagstuhl, Germany, jun 2006.

[KPR04] Kestutis Karčiauskas, Jörg Peters, and Ulrich Reif. Shape Characterization of Subdivision Surfaces: Case Studies. *Comput. Aided Geom. Des.*, 21(6):601–614, July 2004.

[Kra98] Rainer Kraft. *Adaptive und Linear Unabhängige Multilevel B-Splines und ihre Anwendungen.* PhD thesis, Mathematisches Institut A der Universität Stuttgart, 1998.

[KS08] Alexander Konyukhov and Karl Schweizerhof. On the Solvability of Closest Point Projection Procedures in Contact Analysis: Analysis and Solution Strategy for Surfaces of Arbitrary Geometry. *Computer Methods in Applied Mechanics and Engineering*, 197:3045 – 3056, 2008.

[KSV12] Robert Kingdon, Marcelo Santos, and Petr Vaníček. Geoid versus Quasigeoid: a Case of Physics versus Geometry. *Contributions to Geophysics and Geodesy*, 42, 2012.

[Lee05] John M. Lee. *Introduction to Smooth Manifolds*. Springer, 2005.

[Leh09] N. Lehmann. Flächenglättung mittels der eingebetteten Weingartenabbildung. Master's thesis, TU Darmstadt, 2009.

[Lev06] Bruno Levy. Laplace-Beltrami Eigenfunctions Towards an Algorithm That "Understands" Geometry. In *Shape Modeling and Applications, 2006. SMI 2006. IEEE International Conference on*, 2006.

[LLM00] Byung-Gook Lee, Tom Lyche, and Knut Mørken. Some Examples of Quasi-Interpolants Constructed from Local Spline Projectors. In *In Mathematical Methods in CAGD: Oslo 2000, Vanderbilt*, pages 243–252. University Press, 2000.

[LR12] Nicole Lehmann and Ulrich Reif. Notes on the Curvature Tensor. *Graph. Models*, 74(6):321–325, November 2012.

[LYC+06] Tong-Yee Lee, Chih-Yuan Yao, Hung-Kuo Chu, Ming-Jen Tai, and Cheng-Chieh Chen. Generating Genus-n-to-m Mesh Morphing Using Spherical Parameterization: Research Articles. *Comput. Animat. Virtual Worlds*, 17(3-4):433–443, July 2006.

[Mac78] Carl Machover. A Brief, Personal History of Computer Graphics. *Computer*, 11(11):38–45, November 1978.

[Mea80] D.J.R. Meagher. Octree Encoding: a New Technique for the Representation, Manipulation and Display of Arbitrary 3-D Objects by Compute. Technical report, Rensselaer Polytechnic Institute. Image Processing Laboratory., 1980.

[Möß06] Bernhard Mößner. *B-Splines als Finite Elemente*. PhD thesis, TU Darmstadt Fachbereich Mathematik, 2006.

[OBA+03] Yutaka Ohtake, Alexander Belyaev, Marc Alexa, Greg Turk, and Hans-Peter Seidel. Multi-Level Partition of Unity Implicits. *ACM Trans. Graph.*, 22(3):463–470, July 2003.

[OF02] Stanley J. Osher and Ronald P. Fedkiw. *Level Set Methods and Dynamic Implicit Surfaces*. Springer, 2003 edition, 2002.

[Pet95] Jörg Peters. C^1-Surface Splines. *SIAM J. Numer. Anal.*, 32(2):645–666, April 1995.

[PH03] Emil Praun and Hugues Hoppe. Spherical Parametrization and Remeshing. In *ACM SIGGRAPH 2003 Papers*, SIGGRAPH '03, pages 340–349, New York, NY, USA, 2003. ACM.

[Pir12] Cécile Piret. The Orthogonal Gradients Method: A Radial Basis Functions Method for Solving Partial Differential Equations on Arbitrary Surfaces. *J. Comput. Phys.*, 231(14):4662–4675, May 2012.

[PL03] Helmut Pottmann and Stefan Leopoldseder. A Concept for Parametric Surface Fitting which Avoids the Parametrization Problem. *Computer Aided Geometric Design*, 20:343–362, 2003.

[PR08] Jörg Peters and Ulrich Reif, editors. *Subdivision Surfaces.* Springer, 2008.

[Pra09] Jennifer S. M. Prasiswa. *Lokale und Globale Algorithmen zur Approximation mit Erweiterten B-Splines.* PhD thesis, TU Darmstadt Fachbereich Mathematik, 2009.

[PU98] Hartmut Prautzsch and Georg Umlauf. A G^2-Subdivision Algorithm. In *Geometric Modelling, Dagstuhl, Germany, 1996*, pages 217–224, London, UK, UK, 1998. Springer-Verlag.

[Rei07] Ulrich Reif. An Appropriate Geometric Invariant for the C^2-Analysis of Subdivision Surfaces. In *Proceedings of the 12th IMA international conference on Mathematics of surfaces XII*, pages 364–377, Berlin, Heidelberg, 2007. Springer-Verlag.

[Rei12] Ulrich Reif. Polynomial Approximation on Domains Bounded by Diffeomorphic Images of Graphs. *Journal of Approximation Theory*, 164(7):954–970, 2012.

[RM08] Steven J. Ruuth and Barry Merriman. A Simple Embedding Method for Solving Partial Differential Equations on Surfaces. *J. Comput. Phys.*, 227(3):1943–1961, January 2008.

[RP99] Alyn Rockwood and Hwajin Park. Interactive Design of Smooth Genus N Objects over a Single Domain. In *Proceedings of the International Conference on Shape Modeling and Applications*, SMI '99, pages 10–, Washington, DC, USA, 1999. IEEE Computer Society.

[Rus07] Raif M. Rustamov. Laplace-Beltrami Eigenfunctions for Deformation Invariant Shape Representation. In *Proceedings of the fifth Eurographics symposium on Geometry processing*, SGP '07, pages 225–233, Aire-la-Ville, Switzerland, Switzerland, 2007. Eurographics Association.

[SAPM04] John Schreiner, Arul Asirvatham, Emil Praun, and B Surface M. Inter-Surface Mapping. *ACM Transactions on Graphics*, 23:870–877, 2004.

[Sch90] Hermann Amandus Schwarz. Sur une définition erronée de l'aire d'une surface courbe. In *Gesammelte mathematische Abhandlungen*, volume 2, pages "309–311". Springer, 1890.

[Sch46] Isaac J. Schoenberg. Contributions to the Problem of Approximation of Equidistant Data by Analytic Functions, Part A: On the Problem of Smoothing of Graduation, a First Class of Analytic Approximation. *Quarterly of Applied Mathematics*, 4:45–88, 1946.

[Sch81] Larry L. Schumaker. *Spline Functions: Basic Theory*. John Wiley and Sons, 1981.

[Sch05] Volker Schönefeld. Spherical Harmonics, 2005. Available online at `http://citeseerx.ist.psu.edu/viewdoc/download?doi=10.1.1.85.6784&rep=rep1&type=pdf`, August 8, 2013.

[Set96] James A. Sethian. *Level Set Methods*. Cambridge University Press, 1996.

[Sis11] Nada Sissouno. *Multivariate Splineapproximation auf Gebieten*. PhD thesis, TU Darmstadt Fachbereich Mathematik, 2011.

[SXG+09] Marcelo Siqueira, Dianna Xu, Jean Gallier, Luis Gustavo Nonato, Dimas Martínez Morera, and Luiz Velho. Technical Section: A New Construction of Smooth Surfaces from Triangle Meshes Using Parametric Pseudo-Manifolds. *Comput. Graph.*, 33(3):331–340, June 2009.

[SYGS05] Shadi Saba, Irad Yavneh, Craig Gotsman, and Alla Sheffer. Practical Spherical Embedding of Manifold Triangle Meshes. In *SMI*, pages 258–267. IEEE Computer Society, 2005.

[SZBN03] Thomas W. Sederberg, Jianmin Zheng, Almaz Bakenov, and Ahmad Nasri. T-Splines and T-NURCCs. *ACM Trans. Graph.*, 22(3):477–484, July 2003.

[Tau95] Gabriel Taubin. Estimating the Tensor of Curvature of a Surface from a Polyhedral Approximation. In *ICCV'95*, pages 902–907, 1995.

[Tri78] Hans Triebel. *Interpolation Theory, Function Spaces, Differential Operators*. North-Holland Mathematical Library, 1978.

[TRZS04] Holger Theisel, Christian Rössl, Rhaleb Zayer, and Hans-Peter Seidel. Normal Based Estimation of the Curvature Tensor for Triangular Meshes. In *In PG 04: Proceedings of the Computer Graphics and Applications, 12th Pacific Conference on (PG04*, pages 288–297. IEEE Computer Society, 2004.

[Van09] Petr Vaníček. Why Do We Need a Proper Geoid?, 2009. `http://www.fig.net/pub/fig2009/papers/ts03c/ts03c_vanicek_3259.pdf`, April 12, 2013.

[VGJS11] A.-V. Vuong, C. Giannelli, B. Jüttler, and B. Simeon. A Hierarchical Approach to Adaptive Local Refinement in Isogeometric Analysis. *Computer Methods in Applied Mechanics and Engineering*, 200:3554–3567, December 2011.

[Wen05] Holger Wendland. *Scattered Data Approximation.* Cambridge University Press, 2005.

[WHL+07] Hongyu Wang, Ying He, Xin Li, Xianfeng Gu, and Hong Qin. Polycube Splines. In *Proceedings of the 2007 ACM symposium on Solid and physical modeling*, SPM '07, pages 241–251, New York, NY, USA, 2007. ACM.

[WP97] Johannes Wallner and Helmut Pottmann. Spline Orbifolds. *Curves and Surfaces with Applications in CAGD*, 1997.

[ZWH12] Yongjie Zhang, Wenyan Wang, and Thomas JR Hughes. Solid T-spline Construction from Boundary Representations for Genus-Zero Geometry. *Computer Methods in Applied Mechanics and Engineering*, 249:185–197, 2012.

Wissenschaftlicher Werdegang

	Nicole Lehmann
1990-2003	Schulausbildung
2000-2001	Westover School, Connecticut
2003	Abitur am Leibnizgymnasium Wiesbaden
2003-2008	Studium der Mathematik an der Technischen Universität Darmstadt
2008	Diplom in Mathematik mit Nebenfach BWL
2009-2013	Promotionsstudium an der Technischen Universität Darmstadt in der Arbeitsgruppe Geometrie und Approximation